Surfactants in Lipid Chemistry: Recent Synthetic, Physical, and Biodegradative Studies

Edited by

J. H. P. Tyman
Brunel University

ROYAL
SOCIETY OF
CHEMISTRY

The Proceedings of a one day meeting held by the Lipids Group of The Royal Society of Chemistry Perkin Division at Brunel University on 18th September 1991.

Special Publications No. 118

ISBN 0-85186-395-7

A catalogue record for this book is available from the British Library

Published by The Royal Society of Chemistry,
Thomas Graham House, Science Park, Cambridge CB4 4WF

Printed by Bookcraft (Bath) Ltd

Preface

This book originates from a one-day meeting held at Brunel
University in September 1991 entitled 'Surface Active
Lipids'. The title may have been misleading. If Bloor's
definition is adhered to that lipids are hydrophobic, are
related structurally to the fatty acids and usable by
living organisms, then the only surface active types are
the phospholipids and even these contravene the definition
through their water solubility. In fact there are
virtually no naturally occurring surface active lipids.
The vast majority of surface active compounds having
'lipidic' character are semi-synthetic or derived from
fossil fuel sources, typically examples being 'soap' and
alkylbenzene sulphonates. The title of this volume,
'Surfactants in Lipid Chemistry' reflects this view.
The subject of surface active compounds and indeed of
lipids generally remains relatively unknown despite the
plethora of detergent and 'polyunsaturate' advertisements
which assail the populace through television. Nevertheless
there is a growing awareness of the existence of
replenishable natural sources, of the finite nature of
fossil fuel materials and concern regarding environmental
or 'green' problems. This state of understanding exists
although the oils and fats and the petroleum industries are
the largest processors of natural products and fossil fuels
respectively. While the proportion of both these resources
used for surface active purposes is relatively small each
is equally important. The contribution from lipid raw
materials has increased in the past four decades although
the author recalls that at that time research on new
surfactants based on lipid or semi-synthetic lipid sources
was thought irrelevant compared with those of petrochemical
origin. Now both types of raw material are recognised as
essential. In the present circumstances a symposium on
surfactants would need to have representative contributions
from both spheres. Organised by the Lipid Group of Perkin
Division its title was intended to include the whole field
of surfactant substances. The first session was entitled
'Synthesis and Properties' and the second, 'Biodegradation
and Environmental Aspects'. There was inevitably some
overlap between the two sessions with both emphasising
natural products and semi-synthetic as well as purely
synthetic materials. The present volume consisting of a
preliminary section, 'Synthesis and Properties', and a
concluding section, 'Synthesis Structure and
Biodegradation', probably represents the proceedings of the
Symposium more accurately. It could be regarded as
complementary in its approach to 'Industrial Uses of
Surfactants', organised by the Industrial Division of the
RSC in September 1991 and held at Salford University.

In his contribution to the first session, 'Natural Base Surfactants-some aspects of their chemistry', K. Coupland reviews the whole area of nonionic, amphoteric, cationic and anionic surfactants derived essentially from triglycerides. He surveys the semi-synthetic uses of triglycerides, fatty acids, their esters and derived fatty alcohols and amines and this review affords a link with the Salford Symposium through mention of new uses and the next contribution by reference to the 'new' alkyl glycosides, descendants from the classical work on carbohydrates at the end of the last century! E.N. Vulfson in his chapter discusses the strategies involved in 'Enzymatic Synthesis of Surfactants'. The structures involve glycolipids derived from fatty acids and disaccharides, lipidic amino acids and lipidic peptides of semi-synthetic origin. The opportunities for biotechnological approaches emphasise the role of biocatalysts under non-aqueous conditions, the usage of highly thermostable enzyme sources, naturally occurring renewable sources and the possibilities for obtaining low-toxicity and biodegradable products. The discussion of lipidic amino acids and peptides and that of carbohydrate esters gives an entry into the contributions of both T.M. Herrington and I. Toth. The 'Properties of Sucrose Esters', given by T.M. Herrington describes the physical studies made over many years on sucrose esters. Their synthesis, phase diagrams with water and n-decane, the physical properties of their aqueous solutions, the equilibrium film thickness of a model emulsion system and the effect of these esters on the transport of a monocarboxylic acid through an oil membrane are all examined in this account. Reference is made to liquid crystalline properties an area in which a prodigious volume of work has been achieved in many different laboratories outside the lipid field on a wide variety of substances with ensuing technological developments which could not have been imagined a few decades ago. In the final contribution to the first session by L. Thompson on 'Surface Chemistry and the Detergency of Surfactants', fundamental work on the mechanism of detergency was addressed. A correlation between physical properties and organic structure is essential if the synthesis of surfactants is to have a rational basis. Doubtless before a molecular modelling approach could be adopted a detailed knowledge of physical mechanisms is necessary and such a picture is given with admirable clarity in this presentation which goes in detail into just one aspect of the action of an anionic detergent.

In the second session of the Symposium, studies on certain lipidic natural products, biochemical and medicinal aspects, the biodegradation of anionic surfactant and of glycerides were reported. I. Toth in 'Synthesis, Resolution and Structural elucidation of Lipdic Amino acids, their Homo- and Hetero-oligomers and drug conjugates', gave an account illustrating the many-sided character of surface active substances. Lipidic amino acids apart from their use in lubricants, cosmetics, polishes and as surface improvers for ceramics, detergents

and biocompatible coatings could find possible uses in drug
delivery systems and in drug formulation. The influence of
lipidic amino acids in aiding the absorption of certain
penicillins and cephalosporins which although having an
outstanding clinial success are ineffective orally has been
studied in detail by this group. In recent years intense
interest in the environmental fate of chemicals has been
manifested not least in that of surfactants partly through
their potential occurrence in water supplies. G.F. White
in 'Biodegradation of Anionic Surfactants' contributed a
fascinating study of the natural mechanisms operating in
the degradation of alkyl sulphates, dialkyl
sulphosuccinates, alkyl ethoxy sulphates, alkyl sulphonates
and linear alkylbenzene sulphonates. In the case of the
first four groups it seems that microorganisms have evolved
efficient hydrolytic enzymes to enable them to use such
materials for energy sources although a similar situation
does not hold with the last group and their slow
biodegradation is a cause for concern. There are few
naturally occurring sulphonated aromatics to have enabled
microorganisms to evolve enzymes to intereact with this
group. Surfactants in the form of emulsifiers and as
glycerides are important in the food industry and the
contribution by J.L. Kinderlerer, 'Degradation of
Glycerides by Fungi', dealt with this aspect. Glycerides
provide both an energy source for fungi and serve as
substrates for their transformation into simpler chemicals
although free medium chain length fatty acids also inhibit
spore germination. It appears that the conversion of fatty
acids into 2-alkanones, 2-alkanols and esters can be looked
upon as an aspect of rancidity development paralleling the
familiar autoxidative and hydrolytic deterioration of
glycerides. The remaining contribution was concerned with
the chemistry of phenolic glycolipids. D. Minnikin
reviewed 'Mycobacterial Phenolic Glycolipids', which are
effectively O-glycosides of 4-(dimycocerosyl)phthioceryl
phenol, mycocerosic acid being a polymethyl branched long
chain fatty acid and phthiocerol itself comprising a long
chain compound bearing 1,3 diol, methoxy and methyl
branches. These unusual structures, in which the
carbohydrate function is invariably an oligosaccharide,
occur in bovine and human tubercle baccili and in other
animal sources. The main biological activity of the
phenolic glycolipids resides in the oligosaccharide core
while the lipid moiety is thought to provide anchorage of
the 'surface-active' lipid to the mycobacterial cell wall.
Nevertheless stereochemical differences have been found in
the multi-branched lipid chains and these may have
biological significance. The last section of this volume,
'Biodegradable Surfactants derived from Phenolic Lipids',
originally an item in the first draft programme of the
meeting, is concerned with the semi-synthesis of a series
of polyethoxylates from the natural resource Anacardium
occidentale in the form of the industrially-derived
cardanol. The occurrence, extractive separation of
cardanol, and its polyethoxylation are described. The
characterisation of the product, its surface properties and

its biodegradation are examined in relation to other well-known industrial products.

Although the original lectures and the present 'written-up' contributions represent a wide spectrum of interests in the field of surfactants and lipid chemistry perhaps they also serve to illustrate the extent of the chemistry of this group of substances in many adjoining technologies and academic studies. I thank Dr. M.H. Gordon, Secretary of the Lipids Group for his suggestions and cooperation, and the Royal Society of Chemistry (books) for their help. I would like to record that the Symposium proved to be a much appreciated occasion.

LONDON, SW14   JOHN TYMAN

Contents

Natural Base Surfactants — Some Aspects of Their Chemistry and Uses

K. Coupland
CRODA UNIVERSAL LTD, OAK ROAD, CLOUGH ROAD, HULL,
N. HUMBERSIDE HU6 7PH, UK

1. INTRODUCTION

Natural base surfactants are differentiated from natural surface active lipids by being chemically modified lipid derivatives. The natural base is usually a triglyceride ester derived from plant or animal sources. These raw materials provide, upon hydrolysis and physical separation, glycerol and a series of fatty carboxylic acids in the carbon range C8-C22. These materials are valuable building blocks for the commercial manufacture of natural base surfactants. The fatty acids may be saturated : palmitic acid (C16:0), mono-unsaturated : oleic acid (C18:1) or poly-unsaturated : linoleic acid (C18:2).*

Conversion of fatty acids to surfactants draws upon the full capability of synthetic organic chemistry and is performed industrially on a large scale.

These valuable oleochemicals are employed in a large number of applications ranging from cosmetics to explosives.

Naturally occurring surface active lipids are valuable materials and can be used without chemical modification in certain applications. A good example of this class of materials is lecithin. This phospholipid is isolated from the residue remaining after degumming vegetable oils such as soyabean oil. Lecithin is used widely as an emulsifier in foodstuffs and has a small number of other (industrial) applications.

---

*Footnote - The shorthand description gives the carbon number followed by the degree of unsaturation (number of double bonds).

With the trend towards natural ingredients, particularly in personal care products, plant extracts containing saponins, surface active materials with a distinct folkloric appeal, have increased in use. Materials that find commercial use are listed in Table 1.

Table 1   Plant derived Saponins

Plant	Saponin (%)	Use
Quillaja saponaria	quillajasaponin (8-10)	shampoo
Aesculus hippocastanum	escin (8-28)	astringent
Equisetium arvense	equisetonine (5)	astringent
Hedera helix	hedarin	tonic
Panax ginseng	protopanaxadiol protopanaxatriol	conditioner

However, the use of these materials is relatively small when compared to natural base surfactants. Furthermore, their range of utility is somewhat limited.

2. SYNTHESIS OF NATURAL BASE SURFACTANTS

The conversion of triglycerides to surface active materials is an important feature of the oleochemical industry. Various synthetic routes are employed resulting in products that cover the full range of surfactant behaviour ie nonionic, amphoteric, cationic or anionic. Some of these strategies are shown in Table 2.

Table 2 Lipid derived secondary surfactants - Synthesis Routes

- direct from the lipid (triglyceride)
- from a mono- or diglyceride
- from the component fatty acid
- from a fatty acid derivative
  (eg ROH or RNH2)

Conversion of Triglycerides to Surfactants

Natural triglycerides can be reacted with polar reagents (amines, alkanolamines, polyols etc) to give surface active agents with the release of glycerol. This scheme is employed on a limited scale giving rise to products with the same fatty acid distribution

(lipid profile) as the starting triglyceride.   An
example of this route is the production of a
diethanolamide direct from coconut oil (Figure 1).

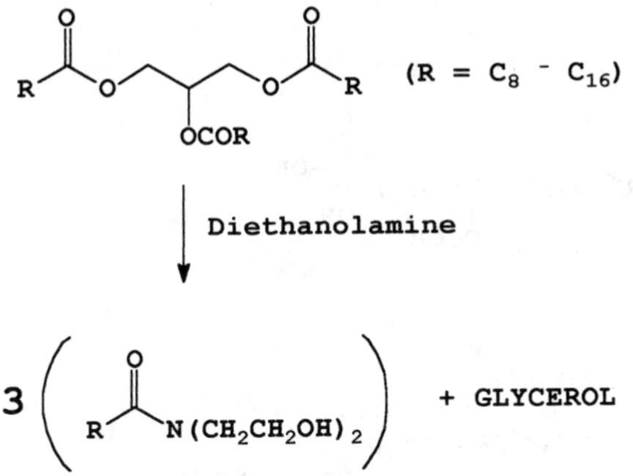

$(R = C_8 - C_{16})$ (1)

Diethanolamine

$3 \left(\begin{array}{c} O \\ \parallel \\ R \quad N(CH_2CH_2OH)_2 \end{array} \right)$ + GLYCEROL   (2)

Figure 1. Cocoalkyl diethanolamide* from coconut oil

Other routes to the same class of materials
involve reaction of the diethanolamine with coconut
fatty acids or methyl esters.   Alkanolamides are used
extensively as foam boosters for anionic surfactants in
shampoos.

Surfactants from Mono-or Di-glycerides

Conversion of a natural triglyceride to a
hemiglyceride by reaction with glycerol gives a mixture
of partial glycerides and unconverted starting
material.   These so-called mono-glycerides are mild
surfactants in their own right and are used in food.
They also can be reacted further to produce more
complex products such as the phospholipid illustrated
in (Figure 2).

In this reaction sequence the starting
triglyceride is soyabean oil which is reacted with
excess glycerol to produce the mixture of partial
glycerides.   For simplicity only the soyabean
diglyceride (3) is shown.   This is reacted with
phosphorous pentoxide, filtered and the phosphoric acid
ester (4) neutralised with ammonia to give a mixture of
ammonium salts (5).

---

*Footnote - The Cosmetics, Toiletries and Fragrance
Association (CTFA) nomenclature for this material is
Cocoamide DEA.   This nomenclature is used elsewhere in
this paper.

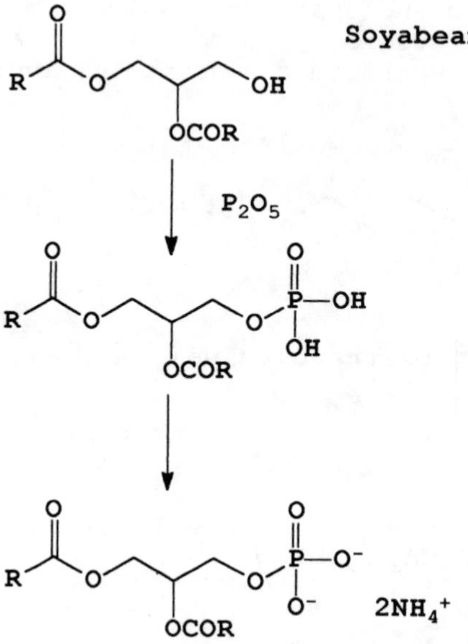

Soyabean diglyceride

(3)

P_2O_5

(4)

$2NH_4^+$

(5)

Figure 2. Phosphate ester surfactant from soyabean hemi-glycerides

The ammonium phosphatide (5) is used as a plasticising surfactant in confectionary and is also useful as a pigment dispersant in cosmetics. The lipid profile of soyabean oil is shown in Table 3.

Table 3 Fatty acid composition of Soyabean oil

Fatty Acid type	carbon number	amount (%)
saturated	16:0	11.0
	18:0	4.0
	20:0	1.0
mono-unsaturated	16:1	0.5
	18:1	22.0
	20:1	1.0
polyunsaturated	18:2	53.0
	18:3	7.5

Both of these examples showing commercial natural base surfactant synthesis give products containing a mixed alkyl hydrophobe (identical with the starting lipid). In many cases this is not a concern and perfectly adequate surfactants result. However, in a large number of cases greater control over the final product is achieved by using an individual fatty acid.

These are isolated from triglycerides using a number of common techniques such as hydrolysis and fractional distillation.

Sorbitan mono-oleate (7) is prepared by the reaction of sorbitol (6) with oleic acid. Oleic acid is obtained from a variety of sources in the oleochemical industry eg animal tallow, olive oil, rapeseed oil and sunflower oil (figure 3).

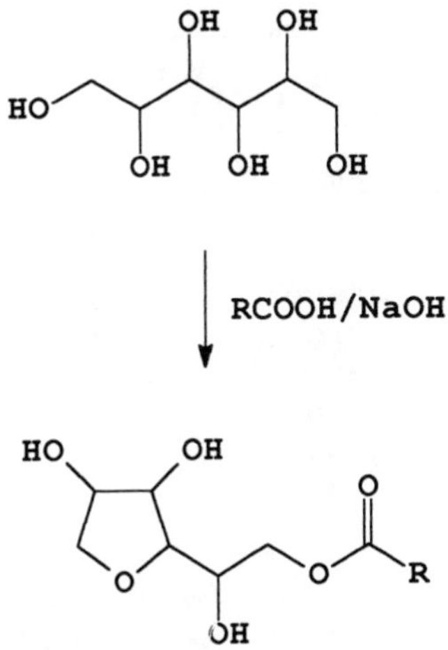

Sorbitol

(6)

Sorbitan mono-oleate

(7)

Figure 3 Preparation of sorbitan mono-oleate

Esterification is accompanied by dehydration reactions resulting in a large number of products. The total mono-esters in commercial products is usually in the range 25-35%. The remaining components being diesters, triesters as well as sorbide esters formed by further dehydration of the sorbitan moeity.

This mixed product is used without further purification and is an excellent emulsifier finding use in a wide variety of applications such as food, cosmetics and explosives (vide infra).

Natural base surfactants from secondary derivatives

Conversion of fatty acids (and fatty acid mixtures) into secondary derivatives, particularly the corresponding alcohol and tertiary dimethylamine, provides many other useful surfactant possibilities. These are summarised in Figure 4.

Secondary Derivative   Surfactant Class

fatty alcohol (RCH$_2$OH)   sulphate esters
                                   phosphate esters
                                   sulphosuccinates
                                   ethoxylates
                                   propoxylates
fatty amines (RCH$_2$NMe$_2$)   quarternary ammonium
                                   salts
                                   betaines
                                   amine oxides

Figure 4.

Natural base surfactants from hydroxy intermediates

In this range of surface active materials the hydrophobe is provided by the hydroxy intermediate. One interesting group of surfactants is the sulphosuccinates. These are esters derived from the reaction of the hydroxy intermediate with maleic anhydride to form a hemimaleate. This activates the double bond in the maleic ester so it can add sodium sulphite.

Three types of hydroxy intermediate are commonly employed in this reaction; fatty alcohols, ethoxylated fatty alcohols and alkanolamides. The reaction is illustrated in Figure 5.

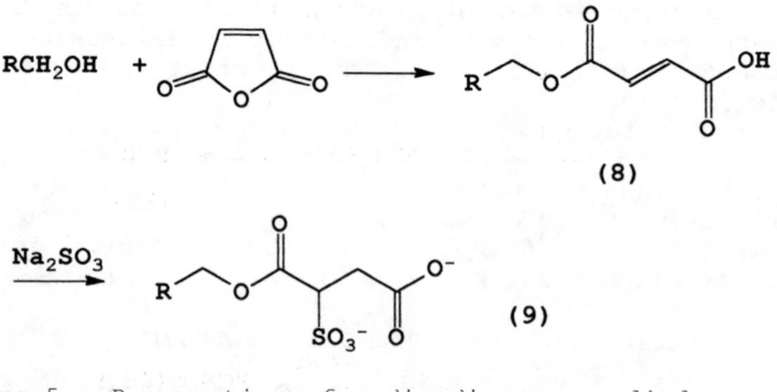

(8)

(9)

Figure 5. Preparation of a disodium mono-alkyl sulphosuccinate

These anionic surfactants are used in shampoo formulations acting as a detergent. They have the advantage of being very mild to skin and eye.

The other two common hydrophobes used in the preparation of sulphosuccinates are ethoxylated alcohols and alkanolamides. Typical structures for surfactants using these intermediates are shown in Figure 6.

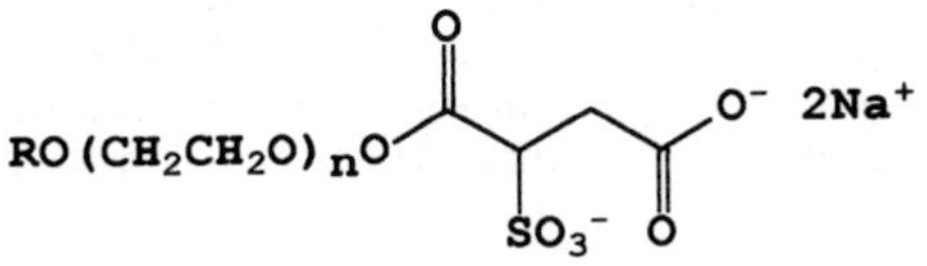

$$RO(CH_2CH_2O)_n \quad \quad \quad O^- \; 2Na^+ \qquad (10)$$

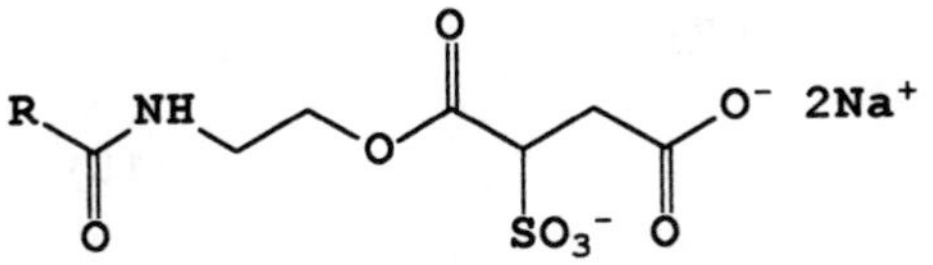

$$R \quad NH \quad \quad \quad O^- \; 2Na^+ \qquad (11)$$

Figure 6. Sulphosuccinates derived from other hydrophobes

Sulphosuccinates are used because of their mildness and not for powerful detergency properties. When used in conjunction with harsh detergents eg sodium lauryl sulphate, they appear to reduce the irritation potential of the sulphate ester.

Conversion of fatty acids to tertiary amines is achieved by reaction with ammonia, dehydrating to the nitrile (12), reduction to a primary amine (13) and final reductive methylation with formaldehyde yields the tertiary amine (14). This scheme is illustrated in Figure 7.

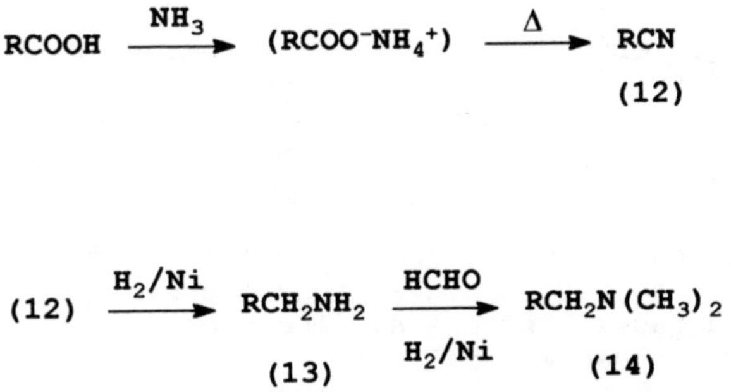

$$RCOOH \xrightarrow{NH_3} (RCOO^-NH_4^+) \xrightarrow{\Delta} RCN$$
$$(12)$$

$$(12) \xrightarrow[]{H_2/Ni} RCH_2NH_2 \xrightarrow[H_2/Ni]{HCHO} RCH_2N(CH_3)_2$$
$$(13) \qquad \qquad (14)$$

Figure 7. Conversion of fatty acids to tertiary amines

Although the process appears complex (and possibly costly) many thousands of tonnes of amine are prepared by this route. The products compete with materials from petrochemical sources ie from alpha-olefins.

An alternative route to give tertiary amines utilises a reactant containing primary and tertiary amino functionality : dimethylaminopropanediamine. Reaction with fatty acids produces the corresponding amide (15) which also has a tertiary amine group (Figure 8).

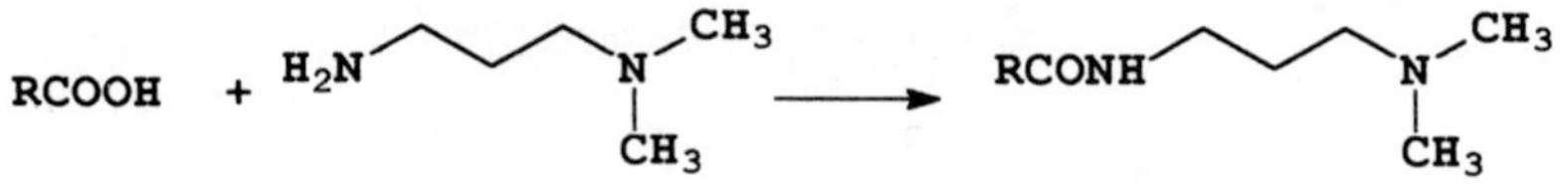

(15)

Figure 8. Preparation of an alkylamidopropyldimethyl amine

Both the intermediates (14) and (15) are used to prepare a number of derivatives which are used commercially as surfactants. Reaction with hydrogen peroxide in aqueous solution produces the amine oxides (16a) and (16b). Amine oxides have been used since the 1940's as foam boosters and stabilisers for anionic detergents in shampoos. They are suspected as being nitrosating agents and may well disappear from use in personal care products.

Reaction with sodium chloroacetate produces betaines, (17a) and (17b). These materials are mild surfactants used primarily as viscosity modifiers for anionic shampoo systems.

The quaternisation of tertiary amines with suitable alkylating agents proceeds rapidly at low temperature. Quaternary ammonium salts (18a) and (18b) are usually produced from methyl chloride.

Quaternary ammonium compounds find wide application as bactericides, fungicides, conditioning agents and fabric softeners. The preparation of these derivatives is shown in Figure 9.

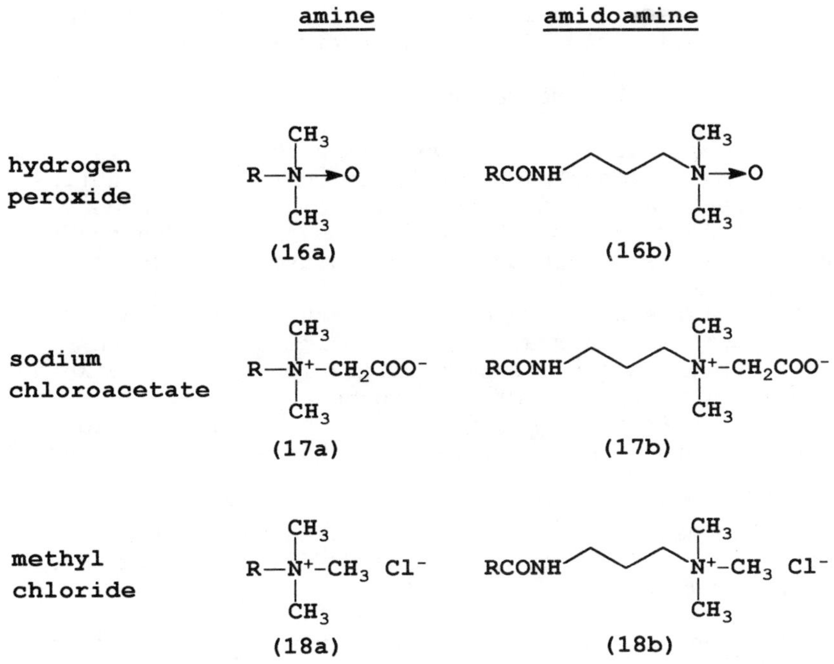

	amine	amidoamine

hydrogen
peroxide

$$R-\underset{\underset{CH_3}{|}}{\overset{\overset{CH_3}{|}}{N}}\rightarrow O$$

(16a)

$$RCONH\diagup\diagdown\underset{\underset{CH_3}{|}}{\overset{\overset{CH_3}{|}}{N}}\rightarrow O$$

(16b)

sodium
chloroacetate

$$R-\underset{\underset{CH_3}{|}}{\overset{\overset{CH_3}{|}}{N^+}}-CH_2COO^-$$

(17a)

$$RCONH\diagup\diagdown\underset{\underset{CH_3}{|}}{\overset{\overset{CH_3}{|}}{N^+}}-CH_2COO^-$$

(17b)

methyl
chloride

$$R-\underset{\underset{CH_3}{|}}{\overset{\overset{CH_3}{|}}{N^+}}-CH_3 \ Cl^-$$

(18a)

$$RCONH\diagup\diagdown\underset{\underset{CH_3}{|}}{\overset{\overset{CH_3}{|}}{N^+}}-CH_3 \ Cl^-$$

(18b)

Figure 9.   Natural base surfactants from tertiary
amines

Amine oxides are described as being quasicationic
as, in acid solution, the nitrogen can be protonated.
They exhibit surprising stability in low pH systems and
have found some use as wetting agents in hypochloride
cleaning compounds.

The betaines have a zwitterion structure at
neutral pH but behave as a cationic salt in acid
solution. Betaines cannot be described as amphoteric
since they cannot give up a proton. This behaviour is
shown in Figure 10.

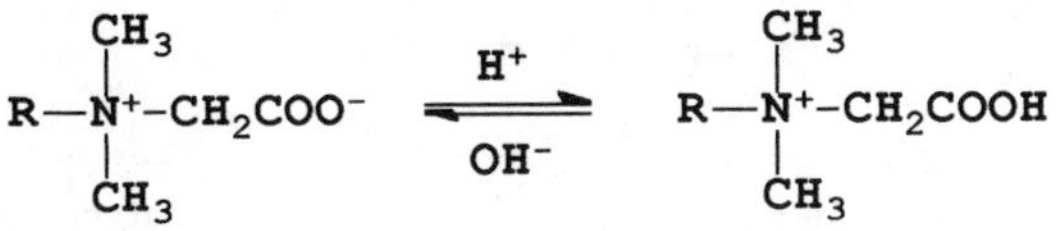

$$R-\underset{\underset{CH_3}{|}}{\overset{\overset{CH_3}{|}}{N^+}}-CH_2COO^- \quad \underset{OH^-}{\overset{H^+}{\rightleftharpoons}} \quad R-\underset{\underset{CH_3}{|}}{\overset{\overset{CH_3}{|}}{N^+}}-CH_2COOH$$

Figure 10.   Effect of pH on the betaine structure

The representative natural base surfactants
produced from nitrogen derivatives of fatty acids
contain acyl groups in the range C8-C22.   The most

popular materials being produced from coconut oil fatty
acids, lauric acid (C12:0) and myristic acid (C14:0).
With increasing molecular weight insolubility in water
can result with some derivatives.

Increasing molecular weight in the tri-methyl
alkyl ammonium salts does however confer some benefits.
These salts are used extensively as after-shampoo
conditioning rinses. The compounds are substantive to
washed hair and once deposited improve the combing
quality of the hair. In addition they dissipate
static charge which gives what is known as "flyaway".
One drawback with quaternary ammonium salts for this
application is the poor eye irritation potential. When
the alkyl group is increased to behenyl (C22.0) eye
irritation is reduced almost to zero. Behenic acid is
obtained from the main fatty acid in certain rapeseed
oils.

The design or selection of surfactants,
particularly, for personal care products, is summarised
in Table 4.

<u>Table 4</u> Selection criteria for commercial surfactants

■   they should be from a "natural",
     "near-natural" and preferably
     vegetal source.

■   they should be biodegradable

■   non-irritating to skin or eyes

■   high foaming

■   cost effective and comply with existing
     legislation.

A great deal of these selection criteria is the
result of a green revolution in the cosmetics and
toiletry industry. It is good for marketing claims and
the use of phrases containing; natural, mild,
biodegradable are often found.

These attributes are sometimes self-exclusive so
the design of a "perfect surfactant" is very difficult.

One class of material which comes close to meeting
these criteria is the sulphobetaines (19) produced by
reacting an alkylamidopropyldimethylamine (15) with
sodium 3-chloro-2-hydroxypropane sulphonate. This
adduct is made by reacting epi-chlorohydrin with sodium

bisulphite. The preparation of these surfactants is shown in Figure 11.

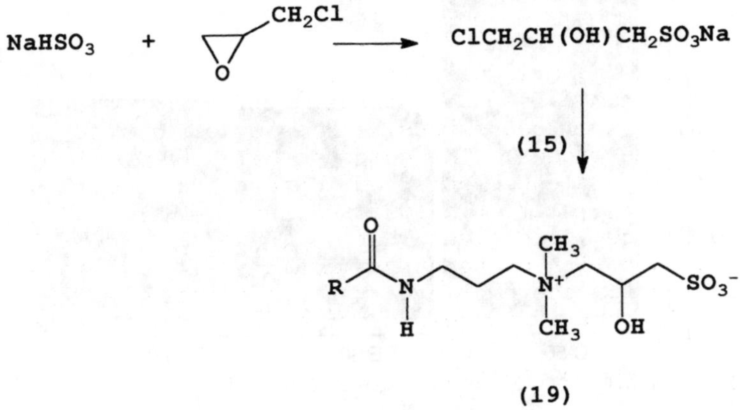

(15)

(19)

<u>Figure 11.</u> Preparation of sulphobetaines

The proposed CTFA name for these surfactants is alkylamidopropyl hydroxy sultaine. Several commercial products exist the most important listed in Table 5.

<u>Table 5</u> Commercially available sulphobetaines

<u>Proposed CTFA*Name</u>	<u>Code</u>	<u>Source</u>	<u>Concentration</u> (%)
Cocoamidopropyl hydroxysultaine	C-50	coconut oil	50
Erucamidopropyl hydroxysultaine	E-30	rapeseed oil	30
Tallowamidopropyl hydroxysultaine	T-30	tallow	30

Sulphobetaines are compatible with other, lower cost, anionic surfactants. Mixtures of these surfactants are extremely mild and are used as the base for baby shampoos, mild shampoos and 'frequent use' shampoos.

To compare the foaming and viscosity characteristics of these materials the materials were compared at equal concentration (5% active ingredient) in combination with sodium lauryl ether sulphate (20%). These results are shown in Figures 12 and 13.

*Footnote - Cosmetics Toiletries and Fragrance Association

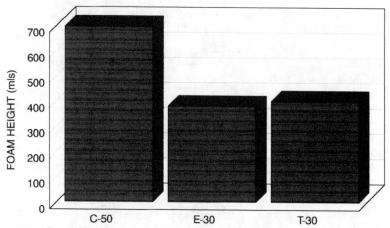

<u>Figure 12.</u> Effect of sulphobetaine structure on foaming

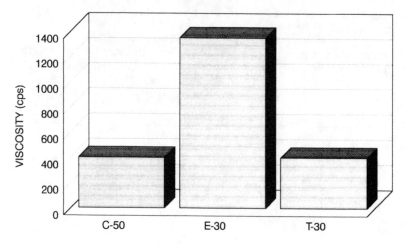

<u>Figure 13.</u> Effect of sulphobetaine structure on
solution viscosity

The surfactant derived from coconut oil (C-50)
gave the greatest foam volume in the Ross Miles foam
test. This test involves rapidly adding the surfactant
solution to a quantity of the same solution in a
graduated cylinder. Foam height is measured initially
(flash foam) and at intervals to judge foam stability.
The two higher molecular weight materials were
approximately equal.

In this latter comparison, somewhat surprisingly,
the erucic acid derived product (E-30) gave the best
result while the two other derivatives were about
equal.

Another important characteristic is the stability of foamed surfactant solutions. Initial foam height is measured (flash foam) and again after 5 minutes. Foam stabilisers or boosters are commonly added to shampoo systems, the most efficient being cocoamidopropylbetaine. A comparison has been made between the effect of this material and the corresponding cocoamidopropylhydroxysultaine (C-50) on the foaming behaviour of sodium lauryl ethersulphate (SLES). These data are shown in Figure 14.

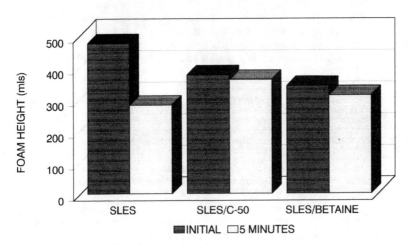

Figure 14. Foam stabilisation by sulphobetaines

These new surfactants have been evaluated in salon tests for 'feel on hair' by volunteers. The overall results have been excellent.

3. DISCUSSION

Natural base surfactants are defined as surface active compounds produced from naturally occurring raw materials. Lipids derived from plant or animal sources (often by-products) are valuable raw materials and capable of extensive modification. It is true that the materials discussed in this paper are not natural surfactants and, in certain circumstances, up to ten chemical reactions may be required to convert the natural raw material into the finished surfactant. Nevertheless, the key raw material, the triglyceride ester, is a renewable resource.

Common oils and fats contain fatty acids having chain lengths in the range C8 - C22. These fatty acids

can be separated into discrete fractions or used as
mixtures. Converting the hydrophobic fatty acid or
ester into a surfactant involves reaction with a polar
intermediate such as a polyol or alkanolamine.
Alternatively the carboxyl group of the fatty acid may
be converted into a secondary derivative such as a
fatty alcohol or amine.

The various transformations possible enable
industry to manufacture the full range of surfactants
including nonionic, amphoteric, cationic and anionic
materials. The end uses for the products are extremely
varied ranging from personal care products to
explosives. In these applications the surface activity
is exploited to provide emulsification properties or
detergency.

The end use for the surfactant clearly dictates
the properties necessary in the material. For example
mildness or compatibility with other components. By
molecular design, through selection of the correct
fatty acyl group or choice of the hydrophilic end
group, a good number of these properties can be
obtained. This has been attempted in the newer
sulphobetaines with promising results.

For the future it is unlikely that any major new
surfactant class will emerge because the cost of
registration for new chemicals has become prohibitively
high. The most recent new introduction of a commercial
surfactant depended on chemistry developed 100 years
ago ! These surfactants are alkyl polyglycosides
prepared by acid catalysed glycosylation of glucose
with fatty alcohols. A 25kt p.a. plant has been built
to produce these "new" surfactants.

The environmental impact of surfactants has also
to be carefully considered with biodegradability being
a major concern. The use of the once major fabric
softener in Europe has ceased because of slow
biodegradability. This product was a natural base
derived cationic distearyl dimethyl ammonium chloride.
The problem material has been largely replaced by
cationic tallow based esters.

4. EMULSIFIERS FOR EXPLOSIVES

At the lipid meeting held at Brunel University the
topic of explosives and the new use of emulsion
explosives was discussed. This took the form of a
series of colour slides showing explosives in action.
Regrettably these photographs cannot be published in
this volume. The thrust of the topic is briefly

summarised below.

The problem of emulsifying saturated ammonium nitrate solutions with a (low phase volume) hydrocarbon was resolved by the use of sorbitan esters eg sorbitan mono-oleate and sorbitan mono-iso-stearate. At moderately high surfactant levels (1-2%) stable w/o type emulsions result. Following sensitisation the explosive mixture can be detonated using conventional detonators.

The key to this system is of course the emulsifier which has to stabilise an extremely sensitive system. Stability is required for periods up to 1 year. The primary emulsion is a transluscent viscous gel which can be pumped. It is this feature which makes this system attractive for large scale open-pit mining.

A difficult problem has been overcome by the use of a natural base emulsifier.

Enzymatic Synthesis of Surfactants

Evgeny N. Vulfson

DEPARTMENT OF BIOTECHNOLOGY AND ENZYMOLOGY, INSTITUTE OF FOOD RESEARCH, EARLEY GATE, WHITEKNIGHTS ROAD, READING RG6 2EF, UK

Introduction

Surface active agents (surfactants) are generally amphiphilic substances consisting of hydrophobic and hydrophilic moieties chemically "linked" to each other in a particular way. According to their structure, surfactants display both hydrophilic and lipophilic properties, the balance between which is dictated by the chemical nature of the constituent parts. The common features of surfactants are their ability to reduce surface/interfacial tension, to disperse and/or solubilize substances which are otherwise insoluble and to affect the foaming properties of aqueous solutions.

Surfactants constitute an important and widely used class of industrial chemicals with a manufacturing volume estimated at tens of millions of tonnes per annum. Due to their wide ranging applications, a variety of individual compounds and blends are currently available on the market. Traditionally, the synthesis of surfactants has been considered solely within the capabilities of organic chemistry. However, the rapid advance in biotechnology over the last decade has led to a considerable growth of interest among industrial companies in "biosurfactants" and larger scale methods for their manufacture[1].

There is also much increased public awareness of the advantages of "green" and "natural" products. As a result, the environmental impact of currently used surfactants is another issue that the manufacturers have had to take seriously into consideration. For example, the biodegradability and toxicity of industrial detergents have already become almost as important as the actual performance of the product. Similarly, more stringent regulations are being introduced in the food industry to minimize the adverse allergic effects of artificial food additives, particularly food emulsifiers.

Although these points certainly justify the effort invested in the development of biological methods for the manufacture of surfactants it should be stressed that biotechnology cannot yet provide an economically viable alternative to the conventional production of the majority of surface active agents that are required. At the same time remarkable progress has been achieved in recent years and it is the author's view that at least some new biotechnological approaches to the synthesis of surface active agents will be commercialised in the near future. Biosurfactants are already competing with the conventional products at the top end of the market (pharmaceutical formulations, ingredients for personal care products such as shampoos, and moisturizing creams) and will probably penetrate the market of food emulsifiers and washing powder detergents before too long.

Generally, there are two alternative strategies that can be adapted for the synthesis of surfactants: whole cell bio-transformations and (chemo-)enzymatic synthesis. The former approach has been the subject of numerous studies resulting in isolation and identification of a variety of suitable microorganisms and the corresponding biosurfactants produced during the fermentation. This topic, however, will receive only a brief consideration in this paper due to the availability of several comprehensive reviews[2-4]. In contrast the application of enzymes to the synthesis of surfactants will be discussed in some detail.

Microbial Synthesis of Surfactants

Various strains of microorganisms and fungi are capable of producing extracelluar surface active lipids. The formation of these lipids is often observed when microorganisms are grown on alkanes. This observation suggests that the physiological function of extracelluar surfactants is to facilitate the utilization of alkanes by emulsification. The majority of extracelluar surfactants described so far, are glycolipids. However, a number of polypeptide-, phospho- and sulfono-lipids have also been described (Fig 1). Typically, the hydrophilic and hydrophobic moieties are connected by ester, amide or glycosidic linkages.

There are several reasons why the microbiological production of surfactants has received so much attention in recent years. First of all the diversity and the structural complexity of biosurfactants (the majority of which are simply unobtainable on a practical scale by conventional chemical methods) have

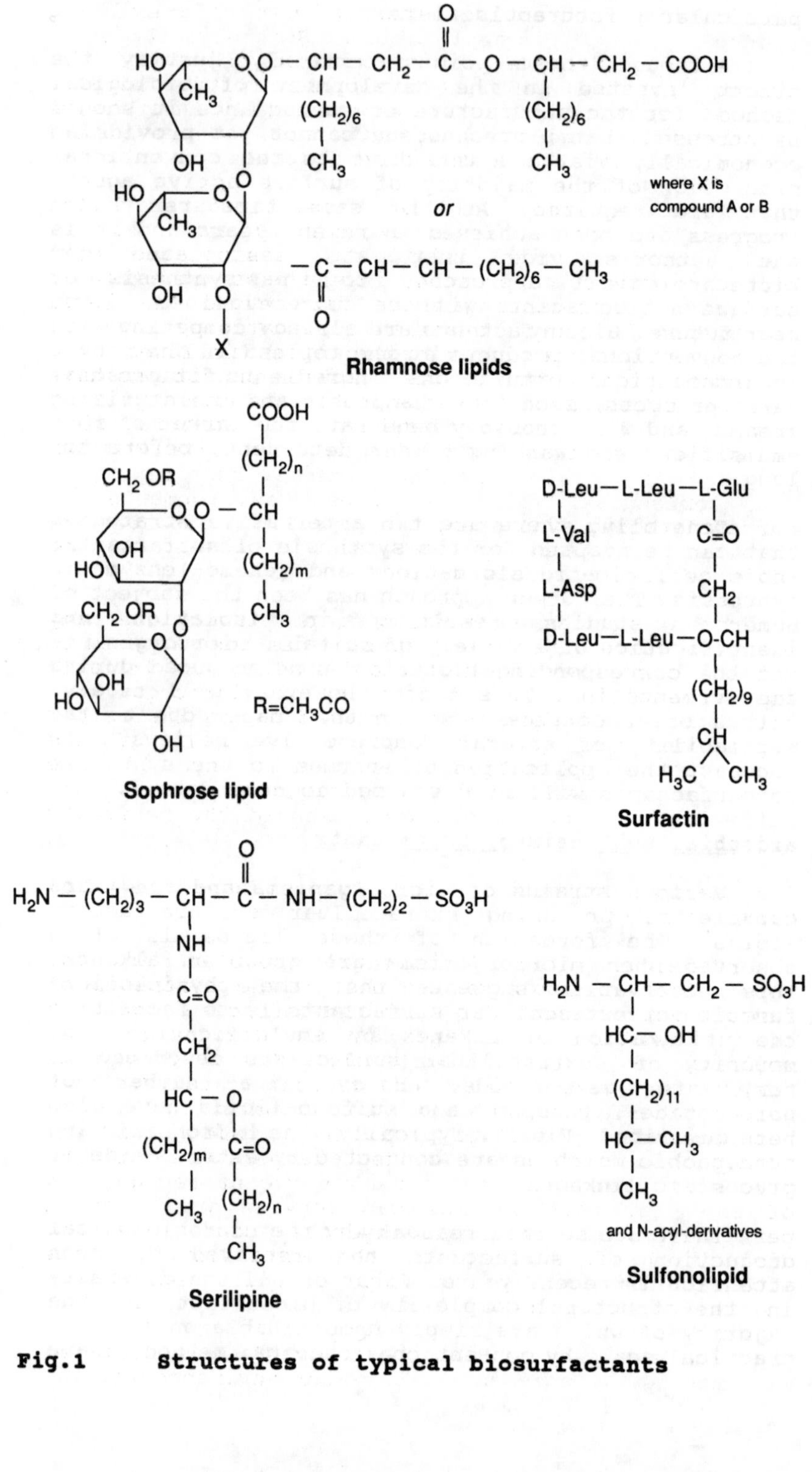
where X is
compound A or B
or
Rhamnose lipids
R=CH3CO
Sophrose lipid
Surfactin
and N-acyl-derivatives
Sulfonolipid
Serilipine

Fig.1 Structures of typical biosurfactants

made them an attractive alternative to the existing products in specific applications. Secondly, there is virtually no environmental concern regarding their use due to rapid and complete biodegradability. Thirdly, fermentation should provide a potentially inexpensive method of large-scale manufacture. The latter, however, appears to be the main obstacle, so far.

Typically the yield of biosurfactants in the fermentation broth rarely exceeds 10 g/l. The low yield combined with difficulties associated with product recovery have often, in the past, resulted in projected manufacturing costs far beyond the level acceptable for many industrial applications. However, intensive screening of microorganisms followed by a thorough optimization of the reaction conditions have led to a substantial increase in biosurfactant yields. For instance, the preparation of sophorolipid[5], succinoyl trehalose-[6] and mannosyl erythritol-lipids[7] at the level of 40-80 g/L has recently been reported. The commercial value of these developments is currently being evaluated; the glycolipid products are expected to find applications in cosmetics and personal care products.

It is also worth mentioning that there are many examples where genetic engineering has been applied to metabolic pathway manipulation leading to dramatic improvements in the productivity of microorganisms. Although this type of approach still remains rather expensive, the generic know-how as well as the knowledge of glycolipid biosynthesis, and in some cases the organization of the corresponding genes[8], are sufficient for the undertaking, should the necessity arise.

Enzymatic Synthesis of Surfactants: General Consideration

Generally the fundamental problem in the application of isolated enzymes to the synthesis of surface active agents has been the "bringing together" of the hydrophilic and hydrophobic domains of what would be a surfactant molecule. For a long time enzymes were perceived as very delicate catalysts that were totally unsuitable for operation in a heterogeneous reaction mixture under extreme conditions, such as elevated temperature or the presence of organic solvents. The recent recognition of the fact that enzymes can survive and function perfectly well in near-anhydrous conditions and display highly enhanced thermostability[9,10], has significantly widened the scope of their application in the production of surfactants.

There are several important advantages associated with the use of enzymes in low-water environments. In

vivo the synthesis and hydrolysis of the same
molecules are always catalysed by different enzymes.
The elimination of water from the medium effects the
"reversal" of hydrolytic reactions and thus avoids the
use of expensive co-factors or activated substrates
required for the "synthetic" enzymes. It is also worth
mentioning that water itself is not an ideal medium
for synthetic purposes since it often participates in
side reactions and may complicate the product
recovery. Additionally hydrolases generally display a
"relaxation" of specificity in anhydrous conditions
and as a result may accept a variety of "unnatural"
substrates. For example, the endopeptidase subtilisin,
the natural function of which is to hydrolyse
proteins, readily catalyses the acylation of sugars
and steroids in organic solvents[11].

Therefore, there is a principal difference
between the microbial and enzymatic synthesis of
surfactants where the type of enzymes involved and
their environment are concerned. The former is a
typical biosynthetic process catalysed by whole cells
and solely dependent on their viability, while the
latter approach is an organic synthesis where enzymes
are used in place of a conventional chemical catalyst.
The two approaches are complementary not only in terms
of the production methods but also because the
different surfactant structures are amenable to both
methodologies.

The synthesis of various surface active agents by
means of hydrolytic enzymes will be considered below.
Although this consideration is mainly centred around
the use of lipases, the potential of other hydrolytic
enzymes such as phospholipases, glycosidases and
proteases will also be discussed in some detail.
Finally, it should be emphasized that practically all
the biotransformations described were actually
performed with bulk industrial enzymes rather than
specially purified preparations.

Synthesis of Monoglycerides

Monoglycerides, as well as their numerous
derivatives (ethoxylated monoglycerides, acetic,
lactic, citric and diacetyl tartaric esters of
monoglycerides), are emulsifiers widely used in the
food industry[12]. Monoglycerides are currently produced
chemically by glycerolysis of fats and oils.
Traditionally, the oleochemical industry operates
glycerolysis almost entirely in a batch mode at
temperatures of 240-260°C primarily to achieve a high
miscibility of the reactants[13]. In addition to the high
energy consumption the resulting products are often
unusable as obtained and require redistillation to
remove impurities and degradation products. Highly
unsaturated, heat sensitive oils cannot be processed

directly without prior hydrogenation. Thus, the conservation of energy and minimization of thermal degradation are probably the major attractions of enzyme-based technology.

Enzymatic glycerolysis of various fats and oils has been performed recently in a nearly stoichiometric solvent-free mixture of substrates at ambient temperatures using a range of 1,3-position specific lipases[14]. The equilibrium shift required for the reaction to proceed towards an accumulation of the final product was achieved by a decrease in the reaction temperature below the melting point of the monoglycerides[15]. Typically, yields of 70-80% were obtained (Table 1).

Table 1

Glycerolysis of Fats and Oils by *Pseudomonas fluorescens* Lipase

Source of fat or oil	Optimum T,°C	Maximum yield of MG,%
Beef tallow	46	76
Lard	30	69
Milk fat	32	80
Palm oil	40	67
Coconut oil	30	77
Rapeseed oil	5	77
Olive oil	10	90
Corn oil	5	42

Optimum temperature is the temperature at which the highest yield of monoglycerides (MG) was obtained (adapted from[15]).

Monoglycerides are also obtainable through enzymatic hydrolysis[16] and alcoholysis[17] of oils catalysed by 1,3-specific lipases. The former method is probably more suitable for the production of mono- and diglyceride mixtures with desirable emulsifying properties whilst the latter enables highly purified monoglyceride products to be obtained. Alcoholysis of oils can be run continuously in a packed column reactor as well as in the batch mode and product separation and recovery are facile. Lipases have also been successfully applied to the production of monoglycerides rich in high value polyunsaturated fatty acids[18,19].

Alternatively, monoglycerides have been prepared in high yields by enzymatic synthesis from glycerol and fatty acids. The reaction has been performed in batch[20,21] and continuously in hollow fibre reactors[22,23]. Extremely high conversion was achieved in batch

experiments when carried out under vacuum to provide
an equilibrium shift towards the final product by the
evaporation of water produced during the course of the
reaction. Similarly polyglycerol fatty acid esters
have been synthesised enzymically in high yields at
ambient temperatures and in a ultimately solvent-free
process (C.J.Kirby & E.N.Vulfson, unpublished
results).

Synthesis of Sugar Fatty Acid Esters

Sugar fatty acid esters are widely used as
industrial detergents and as emulsifiers in a great
variety of food formulations (low-fat spreads,
margarines, sauces, mayonnaises, ice-creams, etc).
Current chemical manufacturing methods typically
suffer from the same drawbacks that apply to the
production of monoglycerides involving high energy
consumption and the formation of undesirable by-
products. Additionally, a whole range of
structurally-similar products are usually obtained due
to the presence of multiple hydroxyl groups in the
carbohydrate substrates. Thus, a typical composition
of sorbitan esters, recently re-examined in the
author's laboratory[24] is depicted in Fig 2. It has been
shown that "sorbitan monolaurate", for example,
consists of at least 65 individual compounds, many of
which were identified by GC-MS as various isomers of
sorbitan, isosorbide and their mono- di- and triesters. (Structures of some possible dehydration
products of sorbitol are shown in Fig 3).

Two main approaches have been pursued so far with
the aim of developing an alternative enzymatic method
for the synthesis of sugar fatty acid esters[*]. The
first was based on the use of organic solvents,
suitable for the solubilization of both substrates[25-30,]
while the second relied upon prior "hydrophobization"
of sugars and their subsequent solvent-free
esterification[31,32]. Although the former approach appears
to be simpler, the reaction kinetics are poor as is
the overall productivity. For this reason the latter
methodology looks more attractive from a technological
standpoint and will be discussed below in some detail.

---

[*]An earlier report on lipase-catalysed esterification
of sucrose and sorbitol in an aqueous solution[33] was
later proven to be unreproducible.

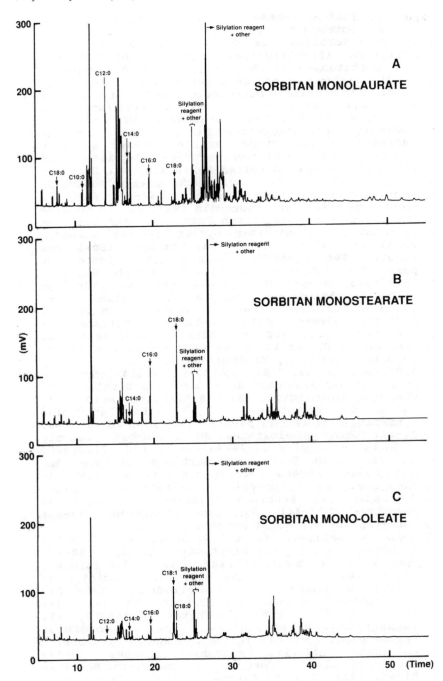

Fig 2. GLC separation of "sorbitan monesters"

Fig 2.   Peak assignment

A.**1:** C8:0(Octaonoic Acid), **2:** C10:0(Decanoic Acid), **3-5:** Isosorbides, **6:** C12:0(Lauric Acid), **7-14:** Sorbitans, **15:** C14:0(Myristic Acid), **16:** Sorbitan, **17:** Sorbitan, **17a:** Sorbitol, **18:** C16:0(Palmitic Acid), **19-21:** Isosorbide Octanoate, **22:** C18:(Oleic Acid), **23:** Sorbitan Octanoate + Oleic Acid + Stearic Acid, **24:** C18:0(Stearic Acid), **25-26:** Isosorbide Decanoate + Sorbitan Octanoate, **27:** Sorbitan Octanoate, **28:** Isosorbide Laurate, **29:** Sorbitan Decanoate + Isosorbide Laurate, **30-34:** Sorbitan Laurate, **35:** Sorbitan Laurate + Isosorbide Myristate, **36-38:** Sorbitan Myristate, **39:** Sorbitan Laurate, **40:** Sorbitan Myristate, **41:** Isosorbide Palmitate + Sorbitan Myristate, **42-43:** Sorbitan Myristate, **44:** Isosorbide Palmitate, **45:** Isosorbide Tetradeceinoate + Sorbitan Palmitate, **46:** Sorbitan Palmitate, **47:** Sorbitan Myristate, **48:** Sorbitan Palmitate, **49:** Sorbitan Palmitate + Isosorbide Oleate, **50:** Isosorbide Oleate, **51:** Isosorbide palmitate - Sorbitan Hexadecenoate, **52:** Sorbitan Palmitate, **53-54:** Sorbitan Oleate, **55:** Sorbitan Palmitate, **56:** Sorbitan Oleate + Isosorbide Oleate, **57:** Sorbitan Oleate, **58:** Isosorbide Oleate **59:-60:** Sorbitan Oleate, **61:** Sorbitan, Dilaurate + Isosorbide Dilaurate, **62:** Sorbitan Dilaurate, **63:** Isosorbide Dilaurate + Sorbitan Dilaurate **64-65:** Sorbitan Dilaurate, **R:** Reagent

B.**1-3:** Isosorbides, **2-8:** Sorbitans, **9:** C14:0(Myristic Acid), **10:** Sorbitan, **11:** Sorbitol, **12:** C16:0(Palmitic Acid), **13:** C18:0(Stearic Acid), **14:** Isosorbide Myristate, **15:** ?, **16:** Isosorbide Myristate 37:) **17:** ?, **18-20:** Isosorbide Palmitates, **21-27:** Sorbitan Palmitates, **28:** Isosorbide Stearate + Sorbitan Palmitate, **29-30:** Isosorbide Stearate, 31: Isosorbide Stearate + Sorbitan Stearate, 32: Isosorbide Stearate + Sorbitan Palmitate + Sorbitan Stearate, **33:** Isosorbide Stearate + Sorbitan Stearate, **34:** Sorbitan Stearate, **35:** Sorbitan Palmitate, **36-44:** Sorbitan Stearates, **45-50:** Sorbitan Di-Esters, **R:** Reagant

C.**1-3:** Isosorbides, **4:** C12:0(Lauric Acid), **5-10:** Sorbitans, **11:** C14:0(Myristic Acid), **12-13:** Sorbitans, **14:** Sorbitol **14a:** ? **15:** C16:0(Palmitic Acid), **16:** ?, **17:** ?, **18:** C18:1(Oleic Acid), **19:** C18:0(Stearic Acid) **20-22:** Isosorbide Myristate, **23:** Isosorbide Hexadecenoate + Sorbitan Myristate, **24:** Isosorbide Palmitate + Sorbitan Myristate, **25:** Isosorbide Hexadecenoate, **26:** Isosorbide Hexadecenoate + Sorbitan Myristate + Isosorbide Palmitate, **27-29:** Isosorbide Oleate **30-32:** Isosorbide Oleate + Sorbitan Oleate, **33-40:** Sorbitan Oleate, **41:** Sorbitan Oleate + Isosorbide Oleate + Diester, **42:** Sorbitan Oleate + Diester, **43-46:** Sorbitan Di-Oleate, **R:** Reagent

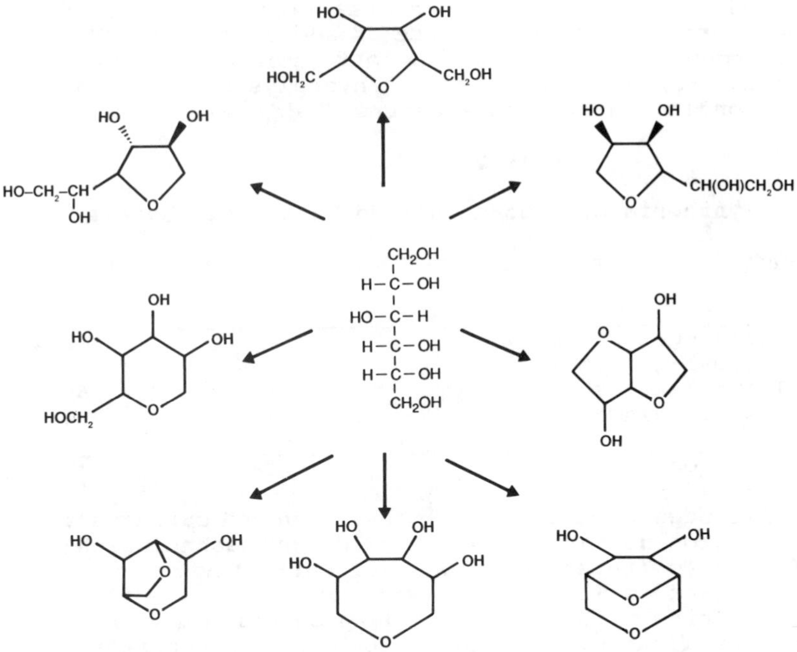

Fig 3. **Possible dehydration products of sorbitol in food grade sorbitan esters.**

Adelhorst et al.[31] have performed a regioselective solvent-free esterification of simple alkyl-glycosides using a slight molar excess of molten fatty acids (Fig.4.1). The rate of the enzymatic esterification was found to be markedly dependent on the length of alkyl group. Thus, only a 20% yield was obtained with glucose and methyl-glucoside after 1 and 21 days of incubation respectively, whilst when ethyl-n-propyl and iso-propyl or butyl-glucosides were used it took only a few hours to complete the reaction. This result clearly indicates that the miscibility of the substrates was not a rate limiting step in the esterification of the latter glucosides. A range of 6-O-acylglucopyranosides were prepared in up to 90% yield and the process has recently entered pilot scale trials. The products are claimed by Novo-Nordisk to be non-toxic, rapidly biodegraded and are expected to find applications as industrial and/or household detergents[34].

A similar process depicted in Fig 4.2, has been recently developed by Fregapane et al.[32] who used monosaccharide acetals as starting materials. As above, prior "hydrophobization" of the sugars was required to carry out the reaction at ambient temperatures and in the absence of added solvents. The final products, monosaccharide fatty acid esters, were obtained after mild acid hydrolysis of the corresponding sugar acetal esters (Table 2).

Table 2

Synthesis of Monosaccharide Fatty Acid Esters

Product	Enzymatic reaction time, h	Yield
6-stearoyl-galactose	16	72
6-palmitoyl-galactose	16	65
5-palmitoyl-xylose	6	50
6-myristoyl-glucose	24	34

Adapted from [35].

Although large scale acetalisation and subsequent deprotection do not seem to present any technological difficulties (these reactions are currently run on a pilot plant scale as a part of the vitamin C manufacturing cycle), the overall production may prove to be complicated. It remains to be seen, therefore, whether these additional steps can be justified by improved quality and the superior emulsifying properties of the final product. At the same time the use of isopropylidene-sugars is probably a more versatile approach when compared to the alkyl-glycosides since it provides an efficient route to the synthesis of monosaccharide fatty esters and can be readily extended to oligosaccharides[35] which may not be amenable to the latter approach.

Synthesis of Amino Acid-Based Surfactants

Amino acid esters and amides are known to be useful as emulsifiers in cosmetic formulations. Apart from excellent emulsifying characteristics, many amino acid-based surfactants have been shown to possess strong antimicrobial activities[36], which also makes them attractive as food additives. Typically, they are prepared by reacting amino acids or their derivatives with fatty acid chlorides or anhydrides (see for example[36,37] and references cited therein).

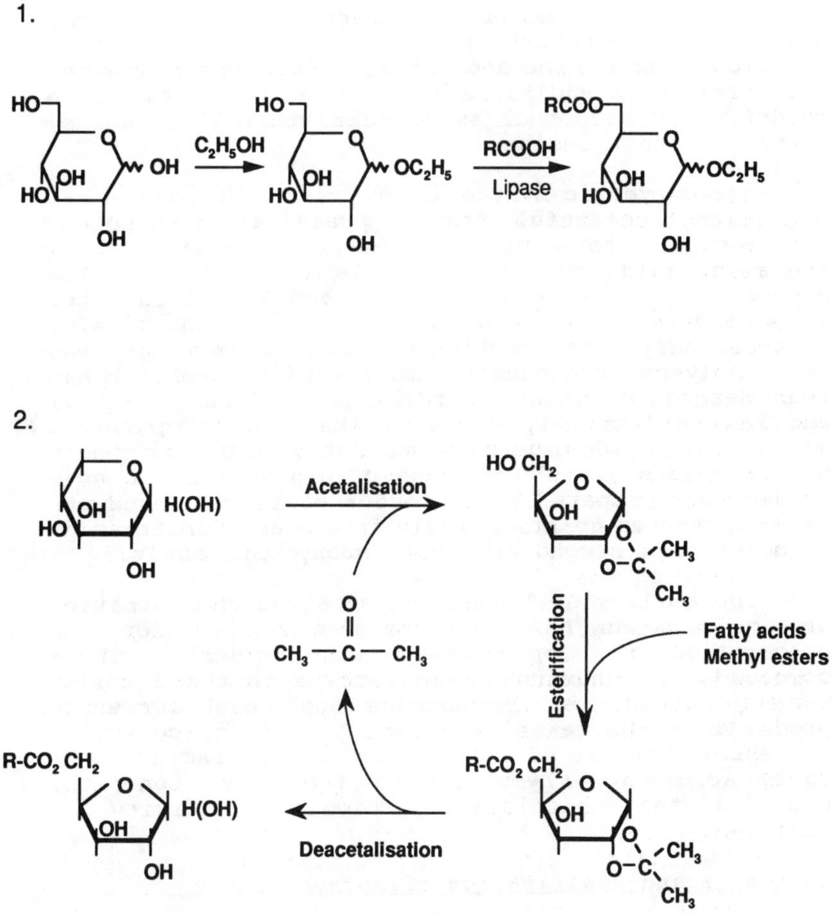

1.

2.

Fig 4. **Lipase-catalysed synthesis of sugar-based surfactants.**

Recently, several attempts have been made to use various lipases in the synthesis of amino acid-based surfactants (Fig 5). Nagao and Kito[38] have prepared O-oleoyl-L-homoserine and carried out a preliminary assessment of its emulsifying properties. The product appeared to be highly efficient in stabilizing oil-in-water emulsions. However, the preparative yield of the synthesis, conducted in a concentrated aqueous solution of the amino and fatty acids, was too low for practical use. Additionally, the authors failed to esterify several other amino acids including serine, threonine and tyrosine.

The enzymatic synthesis of amino acid amides has been more successful from a practical standpoint. Montet et al.[39] have prepared ϵ-N-acyl-L-lysines using the amino acid and vegetable oils as substrates. The products were obtained in 35, 60 and 73% yield after 1, 4 and 7 days incubation at 90°C respectively. Interestingly, the reaction was also shown to proceed under solvent-free conditions. A similar approach has been described by Novo-Nordisk scientists (Fig 5.3) who have selectively acylated the α-amino group of amino acid amides using free fatty acids or their methyl esters[40]. A whole range of N-α-acyl amino acid amides were prepared in up to 50% yield. The products were converted quantitatively into N-acyl-amino acids by means of a second enzyme, carboxypeptidase Y.

The ability of lipases to catalyse the formation of amide bonds has also been exploited for the preparation of simple fatty acid amides[41,42]. These compounds have numerous applications in the textile, paper and polymer industries and are currently produced in relatively high energy-consuming processes. The use of enzymes for the manufacturing of fatty acid amides would offer the above benefits, provided that sufficiently high productivity is achieved.

Synthesis and Modification of Phospholipids

Crude lecithin, produced from plant oils, consists of a complex mixture of individual phospholipids where the major components are phosphotidylcholine and phosphotidylethanolamine[43]. It is used for the production of a wide range of so-called special lecithins, defined as products which have been specifically processed to achieve the required surface active properties. Lecithins have found a variety of applications in the manufacture of paints and leather, in the food industry (bakery goods, chocolate, margarines, instant products) and as a source of individual phospholipids and their derivatives for specific applications in the pharmaceutical industry and in personal care products[44].

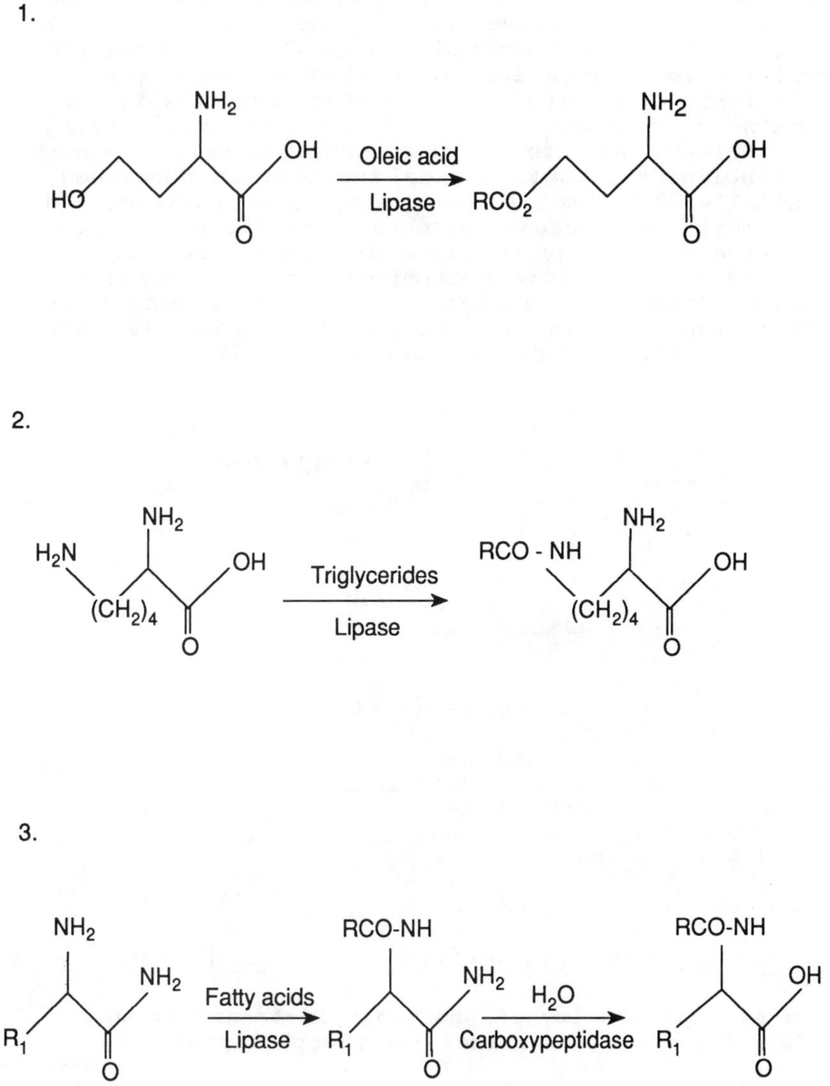

1.

2.

3.

Fig 5. **Enzymatic synthesis of amino acid-based surfactants.**

Although several methods for the chemical and physical modifications of lecithins have been adopted by industry[43], there is also evident scope for the application of enzymes due to the multifunctional nature of phospholipid molecules. Enzymatic modification allows easy control over the degree of hydrolysis with the required regioselectivity, to obtain a product with the desired emulsifying properties[45]. As far as the recovery of the minor phospholipids from the crude lecithin is concerned, typically a complex and expensive purification procedure is needed to obtain products of even moderate purity. At the same time both tasks can be carried out efficiently using commercially available phospholipases and lipases; the sites of attack of these enzymes in a phospholipid molecule are schematically illustrated in Fig 6.

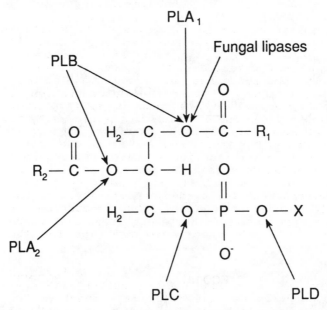

Fig 6. Specificity of phospholipid-transforming enzymes (PL stands for phospholipase)

Thus, phospholipase D has been studied extensively as a catalyst for the synthesis of phospholipids which occur naturally in minor amounts. Phosphatidylserine and phosphatidylglycerol have been prepared in excellent yields from phosphatidylcholine and glycerol or serine respectively[46,47]. Additionally, the enrichment of the major fraction can be achieved by trans-phosphatidylation under low-water conditions with minimal formation of the contaminant phosphatidic acid[48]. Typical results obtained by this method, are summarised in Table 3. Some of these and other similar biotransformations are currently being performed on a multi-kilogram scale by Enzymatix Ltd.

Table 3

Phospholipase D-catalysed Biotransformations of Phospholipids

Substrate, (g/L)		Product yield, %	Reaction time, h
PC	(40)	PC, >95	3
PC	(30)	PG, >97	5
Soybean lecithin	(15)	PC, >95	2
Egg yolk lecithin	(15)	PC, >98	3

Adapted from[46-48]. PC = Phosphatidylcholine
PS = Phosphatidylserine, PG = Phosphatidylglycerol

The complete or partial hydrolysis of lecithins to obtain emulsifiers with improved properties is another example of a current industrial biotransformation. Typically, the batch reaction is catalysed by phospholipase A_2 added to lecithin emulsified in an aqueous buffer at a concentration up to 30% (w/w). However, this process suffers from several complications one of which is the necessity to inactivate phospholipase A_2 after each run since it is practically impossible to recover and reuse the enzyme from the heterogeneous reaction mixture. Irreversible inactivation of phospholipase A_2 is achieved either by a combination of pH-adjustment and heat treatment or by protease treatment followed by inactivation of the less stable proteases[49]. Thus, the production of lysolecithins presents a rather exceptional case where enzyme stability is a major disadvantage for the manufacturers.

Ideally, one would like to run a process continuously in a homogeneous reaction mixture. However, phospholipases are relatively inactive and unstable in lower alcohols which would be a natural choice for phospholipid biotransformations. Another problem associated with the use of phospholipases in such a system, would be the requirement for free Ca^{2+} and a narrow optimum pH range. It would be difficult to maintain these parameters at a constant level under conditions of continuous operation since Ca^{2+} is complexed by the free fatty acids produced during the course of the hydrolysis.

However, fungal lipases are known to function perfectly well in nearly anhydrous alcohols[50] and have recently been shown to accept phospholipids as substrates[51,52]. A highly efficient trans- esterification of lecithins (85-95% yield at the substrate

concentration of 100-200 g/l) has been
demonstrated in a range of primary and secondary
alcohols using a commercially available preparation of
immobilized <u>Mucor miehi</u> lipase (Table 4).

Table 4

Lipase-catalysed Synthesis of Lysophospholipids

Alcohol	Temperature, °C	Reaction time, h	Yield, %
Ethanol	25	8	90
Ethanol + 4% water	25	8	95
Isopropanol	30	16	90
Butanol	37	8	95
Octanol	37	16	85

Adapted from [52].

This biotransformation was run continuously in a
packed column bioreactor for 800 hours without an
appreciable loss of enzymatic activity. The products
of the reaction (sn-1 lyso-lecithin and fatty acid
esters) are easily separated by solvent extraction.
Sn-2 lysolecithin can also be obtained by this process
through intramolecular acyl migration.

Synthesis of Anomerically Pure Alkyl-glycosides

Alkyl-glycosides constitute a promising group of
the new generation of surfactants since they are
prepared from naturally occurring renewable resources
(sugars and fatty alcohols), are easily biodegraded,
and more stable under alkaline conditions than the
corresponding sugar fatty acid esters. Apart from
their usefulness as bulk detergents, anomerically pure
alkyl-glycosides have also found bio-medical and
pharmaceutical applications. Typically, the
preparation of anomerically pure alkyl-glycosides
involves either a multi-step synthesis through
brominated monosaccharide pentaacetates or a
chromatographic separation of the anomers obtained
after direct acid-catalysed coupling[53]. Alternatively,
the latter reaction can be carried out enzymically
using inexpensive and readily available glycosidases
as catalysts.

In a recently developed aqueous-organic two-phase
system, a glycosidase, adsorbed on a polymeric
support, was suspended in a concentrated solution of
sugar that constitutes the aqueous phase whilst a
primary alcohol formed the organic phase[54]. The alkyl-
glycoside, synthesised during the course of the
reaction, accumulated in the organic phase in a
practically pure form. Although the yields were
similar to those obtained by the chemical method, the
anomeric purity of the product exceeded 95%[55].

The main drawback of the enzymatic approach appeared to be the relatively poor kinetics of the reaction, especially when longer chain alcohols were used as substrates. This is probably due to the mass transfer limitations which are bound to exist in such a two-phase system. However, the performance can be improved substantially by carrying out the synthesis in a high surface area hollow fibre bioreactor similar to the ones described for esterification of carbohydrates[22,23], and dipeptide synthesis[56]. The development of a continuous system for the enzymatic production of anomerically pure alkyl-glycosides also looks attractive due to the availability of many highly thermostable glycosidases with diverse specificities

Conclusions

The commercial exploitation of enzymes has traditionally been limited to hydrolytic reactions which are not very useful in the production of surfactants. However, the recent development of biocatalysis in non-conventional media, mainly organic solvents, has provided a new opportunity for bio-organic synthesis. Indeed, enzymes are well recognized at present as a powerful synthetic tool and it is not surprising, therefore, that more and more attention is being paid to the potential use of enzymes in the manufacture of surfactants.

Apart from their high regio- and stereo-selectivity and mild reaction conditions, the ever increasing interest in enzyme-mediated syntheses can be easily explained by two other noticeable trends in the surfactant industry: which are making more extensive use of naturally occurring renewable sources and to minimize environmental and health hazards. Under these circumstances enzymes would be an obvious choice for both utilising natural substrates and for the preparation of low-toxicity, bio-degradable products.

It would be unreasonable to suggest, that the use of biocatalysts in the surfactant industry will shortly become very significant in economic terms. The cost of enzymes still remains one of the major obstacles for the practical realisation of their potential. However recent progress in genetic and protein engineering should enable the enzyme manufacturers to offer recombinant enzymes with superior properties and at a reduced cost[57]. Meanwhile intensive screening programmes have already resulted in the introduction of many highly thermostable biocatalysts that are capable of performing at elevated temperatures for hundreds of hours without any substantial loss of catalytic activity
(see above). This has made the enzymatic processing

cost efficient in the manufacture of many speciality products[34,50,58].

In conclusion I would like to express some hope that this brief review adequately illustrates the prospective and potential benefits in the application of enzymes to the synthesis of surface active agents.

Acknowledgements

The author wishes to thank Prof B.A.Law and Drs M. Whitcombe, J.A. Khan and I.Gill for helpful discussions and suggestions during the preparation of this manuscript. The contributions of G.Fregapane and D.B.Sarney, whose unpublished results were used in this paper, are gratefully acknowledged.

References

1.   J. Hart, Chemistry & Industry, 1989, 6, 384-388

2.   D. Haferburg, R. Hommel, R. Claus and H. Kleber, Adv. Biochem. Engineer. & Biotechnol., 1986, 33, 53-91.

3.   M.I. Van Dyke, H. Lee and J.T. Trevors, Biotechnol. Adv., 1991, 9, 241-252.

4.   G. Georgiou, S. Lin and M.M. Sharma, Biotechnology, 1992, 10, 60-65.

5.   H. Asmer, S. Lang, F. Wagner and V. Wray, J. Am. Oil Chem.Soc., 1988, 65, 1460-1466.

6.   Y. Uchida, S. Misawa, T. Nakahara and T. Tabuchi, Agric. Biol. Chem., 1989, 53, 765-769.

7.   D. Kitamoto, K. Haneishi, T. Nakahara and T. Tabuchi, Agric. Biol. Chem., 1990, 54, 37-40

8.   M.M. Nakano and P. Zuber, CRC Critical Reviews in Biotechnol., 1990, 10, 223-240.

9.   A.M. Klibanov, Trends in Biochem. Sci., 1989, 14, 141-144.

10. J.S. Dordick, Enzyme Microb. Technol., 1989, 11, 194-211.

11. A.M. Klibanov, Acc. Chem. REs., 1990, 23, 114-120.

12. J.B. Lauridsen, J. Am. Oil Chem. Soc., 1976, 53 400-407.

13. N.O.V. Sonntag, *J. Am. Oil Chem. Soc.*, 1984, 61, 229-232.

14. G.P. McNeill and T. Yamane, *J. Am. Oil Chem. Soc.*, 1991, 68, 6-10.

15. G.P. McNeill, S. Shimizu and T. Yamane, *J. Am. Oil Chem. Soc.*, 1991, 68, 1-5.

16. A. Zaks, R. Ivengar and A. Gross, *International Patent Application*, 1991, WO 91/06661 (ENZYTECH Inc).

17. A. Zaks, *International Patent Application*, 1990, WO 90/04033, (ENZYTECH Inc).

18. A. Zaks and A.T. Bross, *International Patent Application*, 1990, WO/90 13656 (ENZYTECH Inc).

19. K. Osada, K. Takahashi and M. Hatano, *J. Am. Oil Chem. Soc.*, 1990, 67, 921-922.

20. S. Yamaguchi, T. Mase and S. Asada, *European Patent Application*, 1986, 0 191 217 (Amano Pharmaceutical Co.).

21. F. Ergan, M. Trani and G. Andre, *Biotechnol. Bioeng.*, 1990, 35, 195-200.

22. M.M. Hoq, T. Yamane, S. Shimizu, T. Funada and S. Ishida, *J. Am. Oil Chem. Soc.*, 1984, 61. 776-781.

23. M.M. Hoq, H. Tagami, T. Yamane and S. Shimizu, *Agric. Biol. Chem.*, 1985, 49, 335-342.

24. G. Fregapane, D.B. Sarney, S.G. Greenberg, D.J. Knight and E.N. Vulfson, In: *Proceedings of Biocatalysis in Non-Conventional Media*, Elsevier, 1992.

25. M. Therisod and A.M. Klibanov, *J. Am. Chem. Soc.*, 1986, 108, 5638-5640

26. S. Riva, J. Chopineau, A.P.G. Kieboom and A. Klibanov, *J. Am. Chem. Soc.*, 1988, 110, 584-585.

27. J. Chopineau, F.D. McCafferty, M. Therisod and A.M. Klibanov, *Biotechnol. Bioeng.*, 1988, 31, 209-214.

28. V. Gotor and R. Pulido, *J. Chem. Soc. Perkin Trans I*, 1991, 491-492.

29. A.E.M. Janssen, A.G. Lefferts and K. van 't Riet, *Biotech. Lett.*, 1990, 12, 711-716.

30.   A.E.M. Janssen, C. Klabbers, M.C.R. Franssen and
      K. van 't Riet, Enzyme Microb. Technol., 1991,
      13, 565-572.

31.   K. Adelhorst, F. Bjorkling, S.E. Godtfresen and
      O. Kirk, Synthesis, 1990, 112-115.

32.   G. Fregapane, D.B. Sarney and E.N. Vulfson,
      Enzyme Microb. Technol., 1991, 13, 796-800.

33.   H. Seino, T. Uchibori, T. Nishitani and S.
      Inamasu, J. Am. Oil Chem. Soc., 1984, 61, 1761-
      1765.

34.   F. Bjorkling, S.E. Godtfresen and O. Kirk, Trends
      Biotech., 1991, 9, 360-363.

35.   G. Fregapane, D.B. Sarney and E.N. Vulfson, J.
      Am. Oil Chem. Soc., 1992, in preparation.

36.   M.R. Infante, J. Molinero, M.R. Julia and P.
      Erra, J. Am. Oil Chem. Soc., 1989, 66, 1835-1839.

37.   S.Y.   Mhaskar,   R.B.N.   Prasad,   &   G.
      Lakshminarayana, J. Am. Oil Chem. Soc., 1990,
      67, 1015-1019

38.   A. Nagao, & M. Kito. J. Am. Oil Chem. Soc.,
      1990, 66, 710- 713

39.   D. Montet, F. Servat, M. Pina, J. Graille, P.
      Galzby, A.Arnaud, H. Ledon, & L. Marcou. J. Am.
      Oil Chem. Soc., 1990, 67, 771-774

40.   S.E. Godtfredsen, & F. Bjorkling. World Patent
      90/14429 (Novo-Nordisk A/S)

41.   D. Montet, M. Pina, J. Graille, G. Renard, & J.
      Grimaud, J. Fat Sci. Technol., 1989, 91, 14-17

42.   R.G. Bastline, A. Bilyk & S.H. Feairheller. J.
      Am. Oil Chem. Soc., 1991, 68, 95-98

43.   C.R. Scholfield. J. Am. Chem. Oil Soc., 1981,
      58, 889-891

44.   W.V. Nieuwenhuyzen. J. Am. Chem. Oil Soc., 1981,
      58, 886-888

45.   S. Fujita & K. Suzuki. J. Am. Oil Chem. Soc.,
      1990, 67, 1008-1014

46.   L.R. Juneja, T. Kazuoka, N. Goto, T. Yamane & S.
      Shimizu.   Biochim. Biophys. Acta, 1989, 1003,
      277-283

47.   L.R. Juneja, N. Hibi, N. Inagaki, T. Yamane & S.
      Shimizu. Enzyme Microb. Technol., 1987, 9, 350-
      354

48. L.R. Juneja, T. Yamane & S. Shimizu. <u>J. Am. Oil Chem. Soc.</u>, 1989, <u>66</u>, 714-717

49. Novo-Nordisk Enzyme Process Division. Lecithase: Product Specification Sheet

50. E.N. Vulfson. In: Lipases: Structure, Biochemistry and Applications, P. Woodley ed., <u>Cambridge Press</u> 1992

51. I. Svennsson, P. Adlercreutz & B. Mattiasson. <u>Appl.Microb. Biotechnol.</u> 1990, <u>33</u>, 255-258

52. D.B. Sarney, G. Fregapane & E.N. Vulfson. <u>Enzyme Microb. Technol.</u>, 1992, submitted

53. S. Matsumura, K. Imai, S. Yoshikawa, K. Kowada & T. Uchibori. <u>J. Am. Oil Chem. Soc.</u>, 1990, <u>67</u>, 996-1001

54. E.N. Vulfson, R. Patel, J.E. Beecher, A.T. Andrews & B.A. Law. <u>Enzyme Microb. Technol.</u>, 1990, <u>12</u>, 950-954

55. E.N. Vulfson, R. Patel & B.A. Law. <u>Biotech. Lett.</u>, 1990, <u>12</u>, 397-402

56. G. Herrman, A. Schwarz, C. Wandrey, M.R. Kula, G. Knaup, K.H. Drauz & H. Berndt. Biotech. <u>Appl. Biochem.</u>, 1991, <u>13</u>, 346-353

57. E.A. Falch. <u>Biotech. Adv.</u>, 1991, <u>9</u>, 643-658

58. J. Casey & A. Macrae, <u>INFORM</u>, 1992, <u>3</u>, 203-207

Properties of Sucrose Esters

T. M. Herrington, B. A. Harvey, B. R. Midmore,
and S. S. Sahi

DEPARTMENT OF CHEMISTRY, UNIVERSITY OF READING, P.O. BOX 224,
WHITEKNIGHTS, READING RG6 2AD, UK

1 INTRODUCTION

Sucrose esters are widely used as emulsifying agents and detergents and have
been on the market for a number of years. The non-toxic nature of the esters
has led to extensive use in the formulation of food products [1]. Their
amphiphilic character can be controlled within wide limits by altering both the
degree of esterification and the chain length of the ester group, so that
extensive permutations are possible to obtain a required hydrophile-lyophile
balance. There has been considerable interest in the interactions of sucrose
esters, since sucrose itself stabilizes the conformation of various proteins in
solution [2,3].

The properties of sucrose esters have been studied over a number of
years in our laboratories. Pure sucrose esters have been synthesised and their
behaviour in aqueous solution and in model emulsion systems has been
investigated. In this article the synthesis of pure sucrose esters, their phase
diagrams with water and n-decane, the physical properties of their aqueous
solutions, the equilibrium film thickness of a model emulsion system and the
effect on the transport of a monocarboxylic acid through an oil membrane will
be covered.

Both thermotropic and lyotropic liquid crystalline properties are
exhibited [4]. In this the sucrose esters resemble the alkyl glycoside and 1-
thioglycosides in forming thermotropic liquid crystalline phases between room
temperature and their melting points and lyotropic liquid crystals in water and
n-decane. The two mono-esters studied, the monooleate and monolaurate, are
both water soluble and form micellar solutions. Both have sharp critical
micelle concentrations and the micellar aggregation numbers were determined
over the temperature range 0 to 70 °C, by freezing point and vapour pressure
techniques. The phase diagrams of sucrose monolaurate, monooleate and
dilaurate in water and of the diester in n-decane were obtained. Fundamental
studies of their emulsion stabilising properties were carried out using laser
interferometry and the equilibrium thickness of the oil-in-water films stabilized
by the monolaurate and monotallowate was determined as a function of
surfactant and indifferent electrolyte concentration. The results, analyzed in

terms of a simple 3-layer (oil-water-oil) model, showed that the repulsive forces increased with added surfactant concentration. In transport experiments it was discovered that sucrose tristearate dissolved in a triglyceride oil, slowed down the rate of transport of valeric acid through the oil membrane separating two aqueous phases.

2 SYNTHESIS

The sucrose esters were prepared using a transesterification procedure from sucrose and methyl laurate or oleate in dimethyl formamide as solvent, by base catalysis. The monoester was separated from sucrose and higher sucrose esters by silica gel column chromatography (eluent petroleum ether/glacial acetic) [5, 6]. As prepared the esters were mixtures of isomers; the sucrose molecule is most readily esterified at the 6, 1' and 6' positions [7]. The purity as estimated by gas and thin-layer chromatography was > 99.5%. From GLC and NMR data on methylated derivatives, the isomers 6': 6: 1' were present in the ratio of 6: 3: 1. The mixed monoester, sucrose monotallowate, was prepared by purifying commercial sucrose tallowate by the same method as above.

3 EXPERIMENTAL METHODS

The phase diagrams were determined using a variety of techniques. Polarising microscopy using a Leitz microscope fitted with a hot-stage was used to identify the phase structures [8]. Lamellar and hexagonal phases were identified by comparison of their textures with photomicrographs in the literature, by the type of conoscopic figure formed and by low-angle X-ray diffraction. Large differences in viscosity and refractive index distinguish the cubic and micellar phases. The transition observed by optical microscopy was confirmed using differential scanning calorimetry. The mixtures were homogenised using a vibromixer and by repeated centrifugation through a narrow constriction. Sealed and stirred samples of known concentration were observed through crossed polarisers, both for heating and cooling cycles in a water bath. The microscopic penetration technique was used to check the phase sequences. Aggregation numbers over the temperature range 0 to 70 °C were determined using freezing point and vapour pressure methods [9]. In dilute solution the osmolality, Θ, is related to the freezing point depression, ΔT, by

$$\Theta = \Delta T/K_f \tag{1}$$

where K_f is the cryoscopic constant, and the resistance changes, ΔR_i, of the thermistors of the vapour pressure osmometer by

$$\Theta = \Sigma_i \zeta_i \Delta R_i \tag{2}$$

where the constants ζ_i are obtained by calibration.

In the interferometry studies, the equilibrium distances between two n-octane droplets in aqueous solutions of the surfactants were measured as a function of surfactant and added electrolyte concentration [10]. The technique depends on the determination of the intensity of light reflected from the two interfaces of a thin oil film sandwiched between oil droplets. The optical arrangement is shown schematically in Figure 1. Light from a helium-neon laser was reflected by an adjustable mirror so that it was incident on the film at a very small angle. The intensity of the reflected light was monitored by a chart recorder as the film thinned. A single layer model was used to calculate the film thickness from the measured intensities, since the paraffin chains of the surfactant are immersed in the oil phase and have almost the same refractive index. For this model

$$\text{Sin}^2 \Phi/2 = \left[\frac{J - J_{min}}{J_{max} - J_{min}} \right] \left[\frac{J_{max} - 1}{J - 1} \right] \tag{3}$$

where the film thickness, h, is related to Φ by

$$\Phi/2 = 2\pi n_1 h/\lambda \tag{4}$$

and $J = I_R/I_0$, the ratio of reflected to incident intensity; n_1 is the refractive index of the film. The intensity of the incident beam, I_0, was determined by measuring the reflection from a small plate of quartz substituted for the film; J_{max} and J_{min} are the last maximum and minimum intensities recorded as the film thins. The use of this formula avoids the necessity of determining the refractive index of the oil phase. Also as the last term in equation 3 is effectively unity, the method is insensitive to the value of I_0.

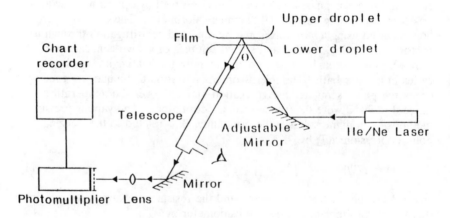

Figure 1 Optical arrangement of the apparatus

Transport measurements were carried out using the Rotating Diffusion Cell. This is designed hydrodynamically so that stationary aqueous diffusion layers of known thickness are created each side of an oil layer [11]. The cell is shown in Figure 2. The oil layer is supported on a porous membrane filter which divides the cell into two compartments, separating the inner and outer aqueous phases. The cell is mounted in a thermostatted glass jacket. The central assembly is rotated by a stepper motor at constant known speeds up to 6 Hz. The fixed slotted baffle, positioned a short distance above the filter, ensures stationary diffusion layers each side of the oil membrane. The 0.06 mm teflon filter, of pore size 200 mm, is located by screwing the stainless steel filter holder against the steel plate attached to the perspex cylinder, leaving an exposed circular area 2 cm in diameter. To measure the flux of an organic acid, the inner compartment, within the perspex cylinder, was filled with the acid and the outer compartment was filled to the same level with 5×10^{-3} mol dm^{-3} NaCl solution. Care was taken to ensure that no air bubbles were trapped beneath the filter and to remove and exclude CO_2 from the solutions by working under nitrogen. The flux of acid passing through the oil membrane was monitored by maintaining the pH in the outer compartment at 7.0 by means of a combined electrode and pH meter attached to a Dosimat automatic titrator which could inject carbon dioxide-free 0.1 mol dm^{-3} NaOH solution.

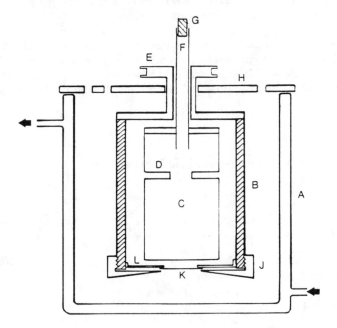

Figure 2 The Rotating Diffusion Cell.
A, thermostatted outer jacket; B, Perspex cylinder; C, PTFE baffle; D, slots; E, pulley; F, stainless steel filling tube; G, rubber bung; H, lid; J, stainless steel filter holder; K, membrane filter; L, stainless steel plate.

4 RESULTS

Phase Behaviour

The thermotropic liquid crystalline transitions of the pure sucrose esters are shown in Table 1 [12]. Both sucrose monooleate and dioleate form gel-like phases at room temperature, whereas the monolaurate and dilaurate are crystalline solids, passing into the gel phase, L_β, on heating. On further heating all four esters form the lamellar phase, L_α. The onset of thermal degradation was shown to be around 180 °C by thermogravimetric analysis, although a darkening in colour occurred between 110 and 120 °C. Layer spacings of 3.7 and 4.3 nm for the lamellar phase of the monolaurate and monooleate respectively were obtained by X-ray diffraction. Sucrose monolaurate is very soluble in water and isotropic micellar solutions are formed with concentrations up to 57 wt%; the solutions showed streaming birefringence at concentrations above 30 wt%. At 20 °C the hexagonal phase, H_1, is formed with increasing concentration of surfactant, followed by the gel phase (Figure 3). Above 87 wt% the lamellar phase is formed on heating. Sucrose monooleate is not so soluble in water as the monolaurate and the phase diagram is shown in Figure 4. The concentrated micellar solutions also showed streaming birefringence, but a region of a viscous isotropic phase, V_1, is interposed between the hexagonal and gel phases. A much larger lamellar phase region is also present.

Considering the diesters, sucrose dilaurate is sparingly soluble in both water and n-decane; the phase diagrams are shown in Figures 5 and 6. In both solvents, regions of lamellar and gel phases are shown, but a small area of a reversed cubic isotropic phase, V_2, is shown only with the hydrocarbon solvent.

Table 1 Transition Temperatures of the Sucrose Esters

	T/ °C		
	Solid - L_β	L_β - L_α	L_α -Isotropic
Sucrose Monolaurate	55	138	163
Sucrose Monooleate	33	94	154
Sucrose Dilaurate	38	81	156
Sucrose Dioleate	-	52	89

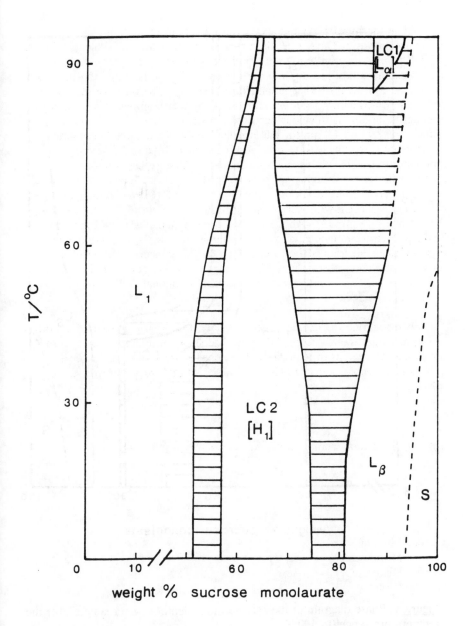

Figure 3 Phase diagram of the sucrose monolaurate + water system over the temperature range 0 - 100 °C. Dotted lines indicate boundaries not determined accurately.

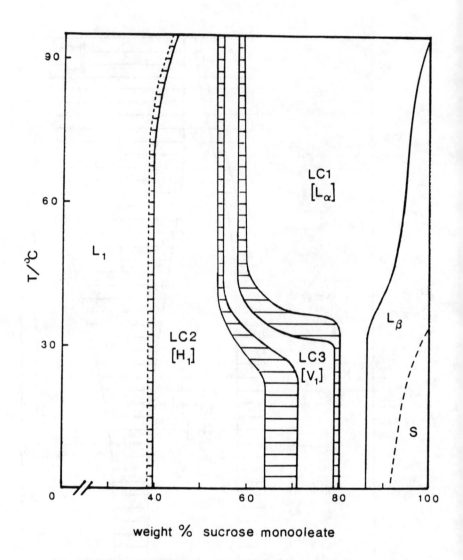

<u>Figure 4</u> Phase diagram of the sucrose monooleate + water system over the
temperature range 0 - 100 °C.

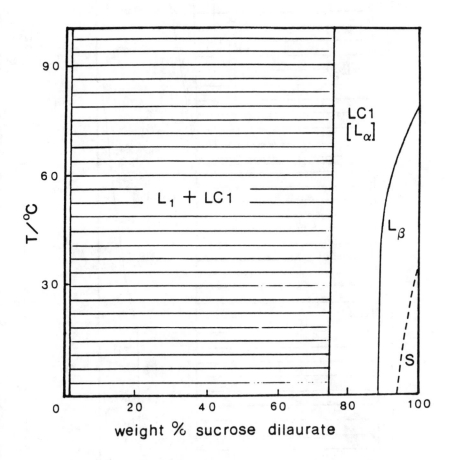

<u>Figure 5</u> Phase diagram of the sucrose dilaurate + water system over the temperature range 0 - 100 °C.

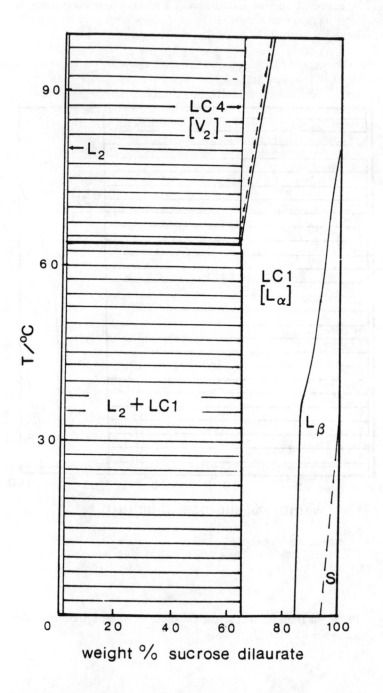

<u>Figure 6</u> Phase diagram of the sucrose dilaurate + n-decane system over
the temperature range 0 - 100 °C.

As might be expected, sucrose dioleate, with a much longer alkyl chain, is very soluble in n-decane as shown in Figure 7. However, only the lamellar and gel phases are formed.

The commercial sucrose monotallowate is a white powder at ambient temperatures with the fatty acid profile: oleic 35%, palmitic 31%, stearic 25% and small amounts of palmitoleic and linoleic acids. Figure 8 shows the phase diagram with water. The compound is only slightly soluble in water below 39 °C, þut the solubility rapidly increases above this temperature to 60 wt%and the lamellar phase is formed. Considerable amounts of diester are suggested by the low solubility, although GLC showed only 5 wt%.

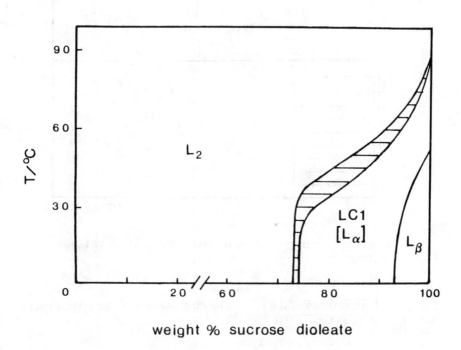

weight % sucrose dioleate

<u>Figure 7</u> Phase diagram of the sucrose dioleate + n-decane system over the temperature range 0 - 100 °C.

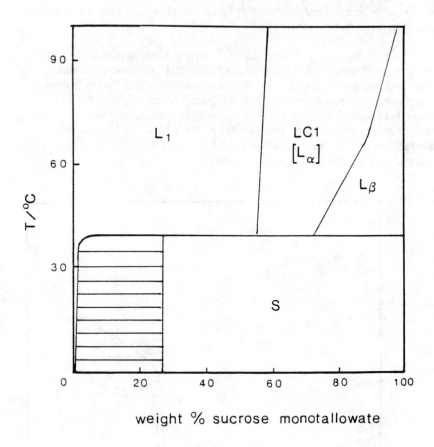

Phase diagram of the sucrose monotallowate + water system over the temperature range 0 - 100 °C.

Micellar Aggregation

The sucrose monolaurate had a cmc of 3.39×10^{-4} mol dm^{-3} at 25 °C in agreement with literature values (4.0×10^{-4} mol dm^{-3} at 25 °C [2]; 3.4×10^{-4} mol dm^{-3} at 27.1 °C [13]). For sucrose monooleate the cmc was 5.13×10^{-4} mol dm^{-3} at 25 °C. Both surfactants had a sharp cmc, so that the thermodynamic data for the solutions were analyzed by assuming that the micelles were monodiperse with a single aggregation number independent of concentration. The single-phase micellar solution was considered to be a 2 component system of solvent and monomer at the cmc together with micelles as solute, so that, by using the theory of McMillan and Mayer [14], the micelle-micelle interactions can be calculated. The osmotic pressure, Π , is given as a power series in the number density, ρ,

$$\Pi / kT = \rho + B_{22}{}^*\rho^2 + B_{222}{}^*\rho^3 + \ldots \quad (5)$$

In this model the non-ideal behaviour is characterized by the virial coefficients $B_{22}{}^*$, etc. In dilute solution, $B_{22}{}^* \approx B_{22}{}^{*0} = - b_{02}{}^0$, the solute-solute cluster integral. The micellar molar mass is assumed to be independent of concentration. The osmotic pressure is related to the water activity, a_1, by

$$\Pi V_1 = - RT\ln a_1 \quad (6)$$

and the osmolality $\Theta = m\phi$ is related to the water activity by

$$\Theta = - \ln a_1/M_1 \quad (7)$$

where M_1 is the molar mass of water. Thus

$$\Theta(M_1/V_1)/c = 1/M_2 + B_{22}{}^*c/M_2{}^2 + \ldots \quad (8)$$

where M_2 is the micellar molar mass, which is assumed to be independent of the concentration c. From equation (8) a plot of $\Theta(M_1/V_1)/c$ against c has an intercept $1/M_2$ and initial slope $B_{22}{}^*/M_2{}^2$. The molality region studied was $0.01 < m < 0.2$ mol kg^{-1}, which is well below the concentration for formation of the liquid crystalline phase. The polynomial fit to the data was found to be linear in c at all temperatures. The values of the aggregation number, η, and second virial coefficient, B_{22} *, are given in Table 2.

Table 2 Aggregation Numbers, η, and McMillan Mayer Virial Coefficients, $B_{22}{}^*$, for Aqueous Solutions of Sucrose Monolaurate and Sucrose Monooleate

T/ °C	\multicolumn{2}{c}{Sucrose Monolaurate}		\multicolumn{2}{c}{Sucrose Monooleate}	
	η	$B_{22}{}^*/\eta/(cm^3\ mol^{-1} \times 10^3)$	η	$B_{22}{}^*/\eta/(cm^3\ mol^{-1} \times 10^3)$
0	51 ± 4	1.47 ± 0.08	97 ± 8	0.877 ± 0.05
25	52 ± 1	1.40 ± 0.04	99 ± 6	0.764 ± 0.04
40	51 ± 2	1.34 ± 0.05	101 ± 9	0.680 ± 0.05
50	50 ± 2	1.30 ± 0.06	104 ± 10	0.616 ± 0.06
60	54 ± 4	1.12 ± 0.08	96 ±13	0.572 ± 0.06

Thin Film Studies

Data were obtained for aqueous solutions of sucrose monolaurate and monotallowate. The concentration of the monotallowate was kept constant at just above the cmc (7.41×10^{-3} mol dm^{-3}) and the concentration of the electrolyte, potassium chloride, was varied between 5×10^{-4} and 7.5×10^{-3} mol dm^{-3}. The results are shown in Figure 9. The equilibrium film thickness decreases as the concentration of the electrolyte is increased at constant surfactant concentration, consistent with decreasing repulsive forces. Two concentrations of surfactant were used for the monolaurate, the cmc and one tenth of the cmc. In each case the film thickness decreased with added electrolyte, but the film thicknesses were greater for the higher surfactant concentration. This shows that increasing the surfactant concentration increases the repulsive force and/or reduces the attractive forces between the n-octane droplets.

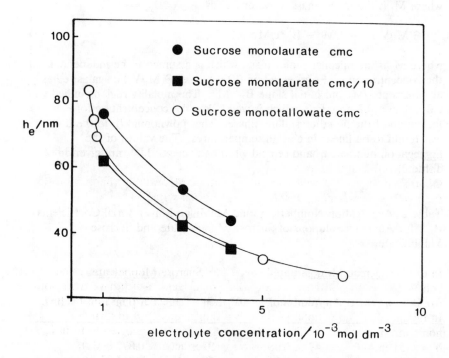

Figure 9 Equilibrium distance (h_e/nm) as a function of the KCl concentration ($c/10^{-3}$ mol dm^{-3}).

Transport of Valeric Acid

The effect of different surfactants dissolved in the oil phase on the transport of valeric acid through triglyceride oil was investigated [15]. A temperature of 45 °C was used to increase the solubility of the surfactants in the oil phase. It was found that a 1 wt% concentration of sucrose tristearate decreased the rate of transport of valeric acid, whereas a 1 wt% concentration of sorbitan trioleate and sorbitan monooleate increased it, the latter quite significantly as shown in Figure 10.

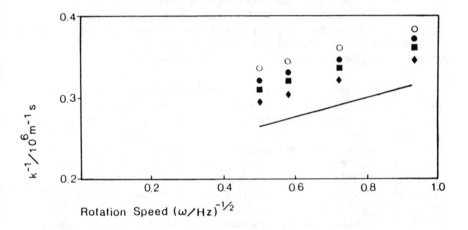

Figure 10 Levich plot of valeric acid through triglyceride ester containing different surfactants at 45 °C.
●, no surfactant; O, 1 wt% sucrose tristearate; ■, 1 wt% sorbitan trioleate; ♦, 1 wt% sorbitan monooleate; ——— Levich slope.

5 DISCUSSION

In displaying thermotropic liquid crystalline behaviour the sucrose esters behave like octyl, nonyl and decyl glucopyranosides [4], which have transitions from the crystal to an intermediate and then a "smectic" phase on heating. In the lamellar and gel phases there is a similar tendency to form homeotropic textures [12]. The dialkyl esters would be predicted to form L_α, V_2 or H_2 mesophases by analogy with ionic surfactants. The dialkyl polyethylene oxide surfactant, $(C_{10}H_{21})_2CHCH_2(OCH_2CH_2)_{10}OH$, in aqueous solution has an extensive L_α phase changing to V_2 above 91 °C. Sucrose dilaurate and dioleate show lower transition temperatures both for L_β to L_α and L_α to isotropic liquid transitions than the monoalkyl esters. Sucrose dilaurate with water shows a more extensive L_α phase than that of the monolaurate, although there is no sign of the hexagonal phase; the V_2 phase is only shown with n-decane. The streaming birefringence exhibited by surfactant-rich isotropic solutions of the monolaurate and monooleate is considered to indicate the presence of rod-shaped micelles and hence the incipient formation of the hexagonal phase. A more extensive hexagonal phase and less lamellar phase is shown by sucrose monolaurate than by the monooleate of longer chain length. In this their behaviour is similar to that of the polyoxyethylene surfactants: for example comparison of $C_{12}EO_6$ with $C_{10}EO_6$ and of $C_{16}EO_8$ with $C_{12}EO_8$ shows that the longer chain length compounds exhibit more lamellar and less hexagonal phase than the latter. For the monoglycerides, however, increasing the alkyl chain length decreases the stability of the lamellar phase [16]. A lower consolute temperature is shown by most of the polyoxyethylene surfactants, implying that the head group hydration decreases with increasing temperature for a given alkyl chain length; this would favour the lamellar phase at higher temperatures in accordance with the phase diagrams. In the temperature range studied the sucrose surfactants show no sign of a cloud point, consistent with the strongly hydrophillic nature of the sucrose head group.

It is possible to use a mass action model to analyze the data and to study the changes in thermodynamic properties on micellization for surfactants with a fairly high and ill-defined value of the cmc, such as octylmethylsulphoxide [17]. However, it is impossible to obtain accurate thermodynamic data in the premicellar region when the cmc is low. The micellar properties of the n-alkyl polyethyleneoxide surfactants, C_nEO_j, which have low values of the cmc, but also show a lower consolute temperature or cloud point, have been studied by neutron scattering, NMR and dynamic light scattering [18-20]. The fundamental issue of, whether the effect of temperature causes predominantly micellar growth or increases the aggregation of small micelles, however, has not been resolved. Sucrose monolaurate and monooleate have very low values of the cmc also, but differ from the C_nEO_j series in two respects; the formation of liquid crystalline phases in the pure state and the absence of a cloud point below 100 °C. At all temperatures studied the aggregation number of the monolaurate is considerably less than that of the monooleate, which can be attributed to the larger hydrocarbon length of the oleyl group. The n-alkylpolyethylene oxide surfactants show similar behaviour; for example at 25 °C the aggregation number, η, is 400 for $C_{12}EO_6$ but 2430 for $C_{16}EO_6$ [9]. Also the η values for the sucrose monoesters are less than those of the C_nEO_j series of comparable chain length and size of headgroup; thus for sucrose monolaurate $\eta = 50$, whereas for $C_{12}EO_6$ $\eta = 400$ at 25 °C. This is consistent with an aggregation number which decreases with the increasing hydrophilic nature of the headgroup. Thus for the sucrose esters smaller micelles are favoured. The lack of a cloud point below 100 °C is explained by the strongly hydrophilic nature of the sucrose headgroup. The previous equation (8) may be written as

$$\Pi / RTc = 1/M_2 + B_{22}*cM_2^2 + \ldots \tag{9}$$

The sucrose monoesters show a consistent trend for the Π /c versus c plots. The initial positive slopes of the curves steadily decrease with increasing temperature, reflecting the steadily decreasing value of $B_{22}*$, the virial coefficient for micelle-micelle interaction. Micelle-micelle attraction with possible aggregation is indicated by a negative value. The decreasing value of the virial coefficient indicates that a cloud point may exist above 100 °C. For $C_{12}EO_6$, B_{22}^* is positive at 18 °C, zero at 20 °C and negative at 45 °C as expected for a cloud point at 50 °C. This is an important difference between the sucrose and polyethylene oxide head groups. A flexible coiled head group is formed by the ethylene oxide chains, a helical coil, which can alter size and shape, whereas effectively a rigid structure is presented by the sucrose head group.

The equilibrium thickness of thin liquid films is determined by the same forces as those determining emulsion stability, so that they are often used as model systems. Emulsion stability depends on 2 processes: the reversible flocculation of the dispersed oil droplets with formation of a thick "Common Black Film"; the further thinning of this film and coalescence of the flocculated droplets in solutions of high ionic strength to form the very thin "Newton Black Film". The thickness of the Common Black Film formed at low ionic strength is determined by the opposition of van der Waals attractive and electrostatic repulsive forces. For the Newton Black Films the van der Waals forces are balanced by steric repulsive forces. Our measurements were restricted to the Common Black Films. In both our work and that of Sonntag et al [21], it was found that for the nonionic surfactant NP20 (nonylphenylpolyethyleneoxide-20) increasing the concentration of the surfactant from one tenth of the cmc to the cmc reduced the equilibrium spacings of the n-octane droplets, which is exactly the same behaviour as found for sucrose monolaurate. However, the reverse behaviour is observed for ionic surfactants [22]. It seems that whether or not the repulsive forces may be increased or reduced by increasing the surfactant concentration is dependent on the system studied. Additionally the values of the film thickness reflect only the equilibrium situation. In a dynamic situation the rate at which the surfactant redistributes itself at the surface as the film thins must play a part. In this connection it was observed that the sucrose esters thinned to a Newton Black Film in a characteristic manner quite different from that of NP20. A black lune formed at the edge of the film and then rapidly spread to cover the whole film for the sucrose esters, whereas for NP20, black spots appeared in the middle which then enlarged to embrace the whole film. This difference must be caused by the different surface viscoelasticities of the films.

The transport data show that sucrose tristearate dissolved in the oil phase inhibits the transfer of valeric acid across a triglyceride oil membrane. This is surprising and certainly shows that no facilitated transport of the small molecule by reverse micelles occurs. The inhibiting effect is probably caused by an increase in the surface viscosity at the oil/water interface. It has been shown that the resistance to transport is greater at the interface than in the oil membrane [15].

The sucrose monoesters of moderately long chain acids are very water soluble and this, together with the increasing distance of closest approach of oil droplets with increasing surfactant concentration, suggests that they should show excellent emulsifying properties. Experiments are underway in this laboratory to determine the rheological properties of the interfacial film, which are undoubtedly a factor in determining long term emulsion stability. The impedance to the transfer of small molecules through an oil phase by the oil-soluble diester indicates a potential application in the preparation of multiple emulsions.

GLOSSARY OF SYMBOLS

a_1	water activity
B	virial coefficient
c	concentration of solute
h	film thickness
J	ratio of reflected to incident intensity
K_f	freezing point depression
k	Boltzmann constant
M_1	molar mass of water
M_2	micellar molar mass
m	molality
n_1	refractive index of the film
R	gas constant
T	absolute temperature
V_1	partial molar volume of water
η	micellar aggregation number
λ	wavelength of helium-neon laser
Θ	osmolality $= m\phi$
ρ	number density of solute
Π	osmotic pressure
Φ	$2\pi n_1/\lambda$
ϕ	osmotic coefficient

REFERENCES

1. T. Yamada, N. Kawase and K. Ogimoto, Yudagaku, 1980, 29, 543.
2. S. Makino, S. Ogimoto and S. Koga, Agric. Biol. Chem., 1983, 47, 319.
3. M. Tomida, Y. Kondo, R. Moriyama, H. Machida and S. Makino, Biochim. Biophys. Acta., 1988, 943, 493.
4. G. A. Jeffrey and S. Battacharjee, Carbohydr. Res., 1983, 115, 53.
5. L. I. Osipow, F. D. Snell, W. C. York and A. Finchler, Ind. Eng. Chem., 1956, 48, 1459.
6. R. K. Gupta, K. James and F. J. Smith, J. Am. Oil Chem. Soc., 1983, 60, 1908.
7. W. C. York, A. Finchler, L. Osipow and F. D. Snell, J.Am. Oil Chem. Soc., 1956, 33, 24.
8. F. B. Rosevear, J. Am. Oil Chem. Soc., 1954, 31, 628.

9.   T. M. Herrington and S. S. Sahi, <u>Colloids and Surfaces</u>, 1986, <u>17</u>, 103.
10.   T. M. Herrington, B. R. Midmore and S. S. Sahi, <u>J. Chem. Soc.,
      Faraday Trans. I</u>, 1982, <u>78</u>, 2711.
11.   A. C. Riddiford, <u>Adv. Electrochem. and Electrochem. Eng.</u>,
      1966, <u>4</u>, 7.
12.   T. M. Herrington and S. S. Sahi, <u>J. Am. Oil Chem. Soc.</u>,
      1988, <u>65</u>, 1677.
13.   I. Osipow, F. D. Snell and J. Hickson, <u>J. Am. Oil Chem. Soc.</u>,
      1959, <u>35</u>, 127.
14.   W.G. McMillan and J. E. Mayer, <u>J. Chem. Phys.</u>, 1945, <u>13</u>, 276.
15.   B. A. Harvey, Thesis, University of Reading, 1992.
16.   E. S. Lutton, <u>J. Am. Oil Chem. Soc.</u>, 1965, <u>42</u>, 1068.
17.   J. M. Corkhill and T. J. Walker, <u>J. Colloid and Interface Science</u>,
      1972, <u>39</u>, 62.
18.   P. G. Neilson, H. Wennerstrom and B. Lindman, <u>J. Phys. Chem.</u>,
      1983, <u>87</u>, 377.
19.   W. Brown, R. Johnsen, P. Stilbs and B. Lindman, <u>J. Phys. Chem.</u>,
      1983, <u>87</u>, 4548.
20.   M. Zulauf and J. P. Rosenbusch, <u>J. Phys. Chem.</u>, 1983, <u>87</u>, 856.
21.   H. Sonntag, J. Netzel and B. Unterberger,
      <u>Spec. Discuss Faraday Soc.</u>, 1970, <u>1</u>, 57.
22.   I. J. Black, Thesis Reading University, 1992.

Surface Chemistry and the Detergency of Surfactants

L. Thompson

UNILEVER RESEARCH PORT SUNLIGHT LABORATORY, QUARRY ROAD EAST, BEBINGTON, WIRRAL, MERSEYSIDE L63 3JW, UK

1   INTRODUCTION

Detergency was defined by Durham[1] as "any procedure for the removal of soil or dirt from the surface of a solid by a liquid". In fact, this general concept covers a diverse collection of soil removal processes, the nature of which vary according to the type and structure of the surface, the nature of the dirt and of the liquid, the hydrodynamic regime and the nature and concentration of the surfactant system. Excluding the hydrodynamic aspects, all of these factors are linked by the fundamental nature of the detergent molecules. All such materials are amphiphilic, containing both hydrophobic and hydrophilic moieties which enable them to form micellar and liquid crystal aggregates in solution and to adsorb at interfaces. It is the process of adsorption with its attendant effects on surface free energy that triggers the various detergency processes.

There are numerous different kinds of surfactant, all of which can be tailored for a particular application by choice of the correct hydrophobe, and in the case of nonionic surfactants, hydrophile (head group) structure. In addition, the properties of surfactants are affected by electrolyte concentration and by temperature. Against this apparently infinite flexibility, the selection of a surfactant system for any particular application is likely to be complicated by a severe set of constraints. For example, surfactants for fabric washing formulations have to be processed in some way depending on whether the final product formulation is to be a liquid, a powder, or a detergent bar. Choice of surfactant is therefore determined partly by the ease of its incorporation into the desired product form. Moreover, in recent years the higher level of environmental awareness in the developed world has become a major factor in the detergent industry's progress toward more biodegradable and less toxic materials. Against these limitations, and taking account of the ever present influence of raw material cost, the problem of maintaining or improving product performance is formidable. Products for fabric washing have to be designed to operate effectively under a variety of product concentrations, temperatures and water hardness conditions. To achieve this requires a thorough understanding of the sensitivity of the various surfactant types to concentration, temperature and electrolyte effects and of the scientific principles of surfactant mixing. This work discusses how far the quantitatively understood principles of surface activity may be applied to the rather more qualitative science of oily soil detergency.

2   EXPERIMENTAL

Materials

Nonionic surfactants, triethylene glycol mono n-dodecyl ether ($C_{12}E_3$) and pentaethylene glycol mono n-dodecyl ether ($C_{12}E_5$) were obtained from Nikko Chemical Company and were in excess of 99% pure. Surface tension versus concentration curves exhibit no minimum, excluding the presence of surface active impurities. Sodium dodecyl sulphate from Sigma was 99% pure, but surface tension drift with time was indicative of the presence of small amounts of dodecanol. Sodium alkyl xylene sulphonate was obtained from Shell, Amsterdam as a 59% aqueous slurry. The mean alkyl chain length was 15.5, consisting of 80% C_{16}. The aromatic group was randomly situated along the alkyl chain such that 90% was internally substituted.

Hexadecane, nominally 99% pure was obtained from Aldrich Chemicals and used after passage through an aluminia column to remove surface active impurities. Squalane and triolein were used as supplied by Sigma. 95% pure triolein was used in detergency experiments, 99% in surface chemical measurements.

Methods

Oily soil detergency was determined by removal, in a Tergotometer, of radio-labelled oil from polyester fabric. The duration of the experiment was 20 min at a rotation speed of 70 rpm. The sensitivity of oil removal to variations in rotation speed in this range is small. For triolein, the label used was ^3H triolein. For hexadecane and squalane it was ^{14}C hexadecane. Where the nature of the oil and its label vary, the assumption is made that fractionation does not occur during the removal process. A soil level of 1.9% w/w on cloth was used throughout and the surfactant concentration was 3×10^{-3} mol dm^{-3}.

Contact angles of oil drops at the polyester/water
interface were made using a surface formed by fusing
the polyester fabric used in the washing experiments
onto clean glass slides. After separation of the
glass and polyester, the polyester surface was found
to be smoother than that of standard Mylar plate,
giving more reproducible contact angles. Oil held in
a 1cm^3 syringe equipped with a micrometer control and
fitted with a cylindrical PTFE tip of diameter 1mm was
contacted with the polyester surface and surfactant
solution carefully introduced so that the oil remained
in contact with both polyester plate and syringe tip.
The polyester plate and the syringe tip were contained
within a 1cm spectrophotometer cell with a light
source behind and a telescope equipped with a
goniometer for angle measurement in front.
Restriction of the path length in this way allowed
accurate assessment of the contact angle even in the
cloudy solutions encountered in some of this work.
After equilibration of temperature the polyester plate
was translated independently of the syringe in a
horizontal plane at right angles to the axis of the
telescope and light source, creating advancing and
receding contact angles. Only the advancing angles
are quoted here. The contact angle was measured
"through the oil phase" so that zero contact angle
corresponds to spreading of the oil.

Oil/water interfacial tensions were generally measured
using a Kruss Spinning Drop apparatus. For tensions
greater than about 3mNm^{-1} the drop volume technique
was preferred.

3   RESULTS AND DISCUSSION

Soil Removal Mechanisms

Relationships between detergency and several physico-
chemical parameters, in particular contact angle and
oil/water interfacial tension, have been reported[2].
These links are quite genuine but are complex in
nature so that apparently simple relationships are
inadequate. Figure 1 demonstrates this, showing, for
a series of four model oily soils, the relationship
between the percentage of oil removed on washing
soiled polyester fabric, and the angle of contact that
a drop of each oil made when contacted with a
polyester plate in a separate experiment. For a
particular surfactant system there is an approximately linear relationship. When the surfactant
system is changed there is still a linear relationship
but it is a different one, so that no general
correlation exists. Similarly, it will be
demonstrated that the commonly accepted relationship
between oil/water interfacial tension (γ_{ow}) and oily
soil detergency is not general either.

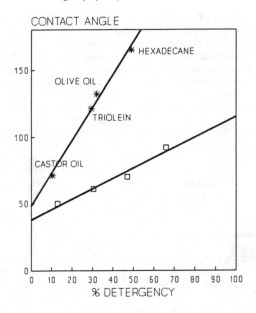

Figure 1. The relationship between oil detergency from polyester fabric and the contact angle of an oil drop on a polyester plate at 25°C. *, 3×10^{-3} mol dm^{-3} SDS/0.1 mol dm^{-3} NaCl; □, 50/50 (mol %) SDS/C$_{12}$E$_3$ at total concentration 3×10^{-3} mol dm^{-3}/0.1 mol dm^{-3} NaCl.

The reasons for the confusion are not hard to identify. They originate with the variable contributions to the overall removal process of multiple removal mechanisms which depend on different combinations of surface chemical parameters. Thus in Figure 1, contact angle is less important to removal by the mixture of sodium dodecyl sulphate (SDS)/C$_{12}$E$_3$ than by SDS alone. It will in fact be shown that γ_{ow} simultaneously becomes more important.

Here, the aim is to expand these mechanistic issues in the context of oil detergency, using a combination of contact angle (θ) and γ_{ow} data to discriminate between the two major oil removal mechanisms, generally referred to as "roll-up" and "emulsification". In the roll-up mechanism (Fig 2a) the balance of surface energies of the oil/fabric/water contact line of an oil film or an oil drop on the fabric is altered by the addition of surfactant, causing the contact angle in the oil phase to increase. Detachment of the drop can then take place either completely, or partially

through the emulsification mechanism (Fig 2b).
Complete removal occurs spontaneously if a contact
angle of 180° is achieved but otherwise a contribution
from the hydrodynamic forces generated in the
surrounding solution is needed. Complete rather than
partial drop removal tends to occur when θ > 90°.

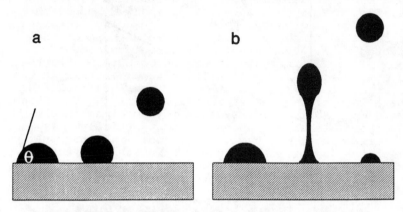

Figure 2. Schematic representation of oil
detergency mechanisms. (a) Complete drop
detachment by "roll-up". (b) Partial drop
detachment by "emulsification"

Partial drop detachment or emulsification occurs when
the surface tension forces holding the drop together
are exceeded by the gravitational and hydrodynamic
forces pulling it apart. The oil/water interfacial
tension, the contact angle and the drop size are all
important in determining the ease with which the drop
ruptures, and the proportion of the drop that
detaches. It is clear that the two removal mechanisms
are not entirely independent of each other and the
task of disentangling them sufficiently to formulate a
general relationship between oil detergency and
measurable physico-chemical parameters is at the least
a formidable one. The surface chemistry of the
oil/water interface has however, been extensively
researched and in the next sections fundamental
understanding of this area is reviewed and a current
evaluation of its links with the process of oil
removal from fabrics is given.

General Principles of Detergency Maximisation

Ethoxylated nonionics are primarily temperature
sensitive, giving rise to o/w interfacial tension
minima which may be "ultra low". Such minima are
associated with the temperature range for
microemulsion formation and the phase inversion
temperature (PIT) for w/o to o/w emulsions[3].
Anionic surfactants are, by contrast, electrolyte
sensitive so that microemulsion formation, phase

inversion, and o/w interfacial tension minima are
associated with salinity[4]. In both cases this
behaviour is understood in terms of molecular
geometry[5]. For nonionics, the effective headgroup
size is controlled by its hydration. Thus a
surfactant having a large headgroup ($a_h > a_c$, see Fig 3)
forms oil-in-water emulsions by way of the natural
curvature that such a molecule provides. Temperature
increase dehydrates the headgroup, reducing its size
until the headgroup area is the same as that of the
solvated tail, when interfacial curvature becomes
zero. This is the phase inversion condition. Further
reduction in headgroup size leads to a reversal in the
direction of curvature at the o/w interface resulting
in the formation of water in oil emulsions. The phase
inversion condition corresponds to a minimum in o/w
interfacial tension. Despite its essentially
empirical nature the geometrical model is a highly
convenient way of picturing phase inversion which is
not too inaccurate provided we think in terms of
"effective" rather than absolute headgroup size.

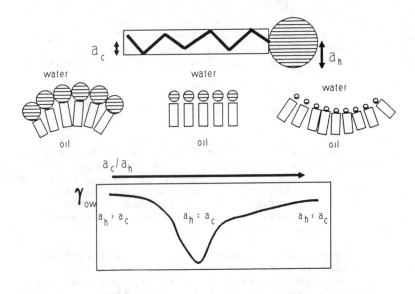

Figure 3. Molecular geometry and the
oil/water interface.

These effects have been discussed thermodynamically by
Aveyard et al.[3] who showed that the interfacial
tension minimum associated with phase inversion for
nonionic surfactants can be viewed as the condition at
which the entropies of micelle and surface formation
are similar. Thus "similarity" of micelle/water and
oil/water interfaces leads to energy minimisation.

The situation for the electrolyte dependence of
anionic surfactant properties is entirely analogous.
Here, the effective size of the headgroups becomes
smaller when their electrostatic interactions are
screened by increasing counter ion concentration. In
this case, the thermodynamic analysis shows that the
interfacial tension minimum arises when the degree of
dissociation of surfactant, into surfactant ion and
counter ion, in the micelle and at the interface are
equal (and low)[4]. This occurs when the micelle
surface is effectively planar. Once again, an
equivalence of micelle and surface leads to a minimum
in energy.

For surfactant mixtures, interfacial tension minima
can be generated when one component of the mixture has
$a_h > a_c$ and the other $a_c > a_h$. At the composition
which gives the average $a_c = a_h$ at the o/w interface,
a minimum in γ_{ow} occurs. The rule of surface and
micellar equivalence holds here also. It has been
shown experimentally by Aveyard[6] for Aerosol OT/
sodium dodecyl sulphate mixtures, and theoretically by
Rosen and Murphy[7] that at the minimum in γ_{ow}, surface
and micellar compositions are equal. Moreover
Holland[8,9] has extended the Regular Solution Theory
of mixed surfactants to relate the lowering of interfacial tension to the mole fraction of the mixture so
that

$$x^s \exp[\beta^s(1-x^s)^2] = x^m[\beta^m(1-x^m)^2]/\exp(\Delta\gamma A_1/kT)$$

where β^m and β^s are the interaction parameters for the
micelle for the planar monolayer respectively, $\Delta\gamma=\gamma_1-\gamma$
where γ is the interfacial tension of the mixture and
γ_1 is that of surfactant 1. x^s and x^m are the mole
fractions of component 1 in the surface and the
micelle respectively. A_1 is the molecular area of
component 1 in a pure monolayer.

In the removal of oil from fabrics or hard surfaces,
the situation is more complex in that the solid/water
and solid/oil interfaces may be as important as the
o/w interface, and the conditions for surface energy
minimisation are different for the various surfaces.
Nevertheless, the same molecular geometry-based rules
of temperature sensitivity for nonionics, electrolyte
sensitivity for anionics and composition dependence
for mixtures apply. What remains to be established is
the extent to which the surface chemical and molecular
geometry based understanding of the o/w interface can
be used to predict oil removal. The literature has
concentrated mainly on temperature/nonionic effects
and here, a link between the phase inversion
temperature (PIT) and the temperature for maximum
detergency, D^t_{max}, has been established by several

groups (PIT is synonymous with γ_{min}). This work is discussed in detail in the next section. In addition, an appreciation of the role of molecular geometry in oil detergency has been shown by Lindman et al.[10] who investigated oil removal from flat surfaces by nonionic and anionic surfactants. They interpret their results in terms of the phase behaviour of the surfactants, involving the critical packing parameter (CPP), where

$$CPP = \frac{V}{la_h}$$

V is the volume of the hydrocarbon chain, l its critical length, generally considered to be close to the extended length of the highest alkyl chain, and a_h is the area of the head group. The CPP concept has been used extensively to define the molecular structures corresponding to the formation of various liquid crystal forms[11] so that spherical micelles require CPP < 1/3 and lamellar phase requires CPP ~1. Lindman suggested that detergency maxima would occur at CPP = 1. By this however, it was not implied that D^t_{max} would always coincide with the onset of lamellar phase formation because the parameter V is a notional quantity with a value subject to the degree of penetration by oil when the surfactant is adsorbed at an o/w interface or to the presence of a co-surfactant in a mixed micelle. Lindman's requirement for CPP = 1 is therefore only another way of defining the phase inversion condition normally characterised by γ_{min}.

The objects of the present work were to test the generality of the emerging principle that phase inversion conditions correspond to optimum detergency and to consider the basis of the detergency maxima in terms of the mechanisms of the oil detachment processes. Most of the work aimed to contrast the behaviour of the non polar oil hexadecane with the relatively polar triolein, although some experiments were done with squalane because it requires significantly different conditions to induce minimum interfacial tension and is therefore a useful diagnostic tool. Phase inversion is brought about in three ways. Firstly using a nonionic surfactant, usually C12E5, inducing surface activity change with temperature variation; secondly with the anionic surfactant sodium hexadecyl σ-xylene sulphonate (C16OXS) by using salinity variation to bring about surface activity change and finally using compositional changes with a mixture of SDS and C12E3, to bring about phase inversion.

The Effect of Temperature on Nonionic Surfactants

The temperature sensitivity of oil removal from
fabrics by nonionic surfactants is well known. Zweig
et al.[12] reported maxima in the removal of mineral
oil from polyester/cotton fabric by $C_{12}E_4$ and $C_{12}E_5$ at
around 30°C and 50°C respectively. Azemar et al.[13]
and also Schambil and Schwuger[14] made similar
observations with hexadecane detergency, again using
$C_{12}E_4$ and $C_{12}E_5$. Both groups related the detergency
maxima (D^t_{max}) to the phase inversion temperature (PIT)
of the appropriate oil/water/surfactant system. Raney
and Miller[15] investigated mixtures of surfactants and
also surfactants with lipophilic additives showing
that both PIT and D^t_{max} vary according to the
surfactant/additive composition. In a further
publication Raney, Benton and Miller[16] again observed
the correlation between maximum detergency and PIT,
reporting correlations at similar temperatures to
those recorded by Azemar and by Schambil and Schwuger.
They also extend this general principle to mixtures of
oils, finding PIT and D^t_{max} for a 50/50 hexadecane/
squalane mixture around 40°C for C12E4 and at 60°C for
C12E5. Thus both PIT and D^t_{max} are dependent on the
composition of the oil.

In a more detailed study along similar lines, Mori,
Lim, Raney, Elsik and Miller[17] also show correlation
of D^t_{max} with PIT for the triglyceride oil triolein,
and also for triolein/hexadecane mixtures. Their
results for the mixture are a little surprising in
that the temperature profile of detergency for the
50/50 mixture is more or less the same as that for
triolein alone. This appears to contradict the
principle suggested in the earlier paper[16] that the
PIT of a mixture of oils is roughly proportional to
their volume fraction and that PIT corresponds to the
maximum in detergency. The data in the present work,
which are reported below, broadly substantiate the
earlier findings with mineral oils. However they
differ in several ways from the results of Miller
et al.[17] for triolein.

Figures 4 and 5 show detergency and interfacial
tension versus temperature plots for $C_{12}E_4$ and $C_{12}E_5$
which are in full agreement with the earlier work in
that they demonstrate a coincidence of the published
PIT with the interfacial tension minimum and D_{max}
measured here. In addition to this however, the data
for $C_{12}E_5$ show the presence of a separate peak at
lower temperature. The presence of this peak is
completely reproducible and its existence implies that
two oil removal mechanisms, with different optimum
temperatures are operating. Figures 6 and 7 show
similar data for triolein, demonstrating that D^t_{max} and
γ_{min} coincide in the region of 25 and 50°C for $C_{12}E_4$
and $C_{12}E_5$ respectively. Again there is some

suggestion of a low temperature peak for $C_{12}E_5$, though it is not clearly resolved. Neither the PIT nor the detergency data correspond to the findings of Miller et al. who observe high detergency at temperatures up to 50°C for $C_{12}E_4$ and 70°C for $C_{12}E_5$, quoting an "apparent PIT" of 68°C for $C_{12}E_5$/triolein. It is not possible to define the reasons for this discrepancy with the detergency data but the difference in the PIT data can be explained. The PIT data of Miller et al. were obtained by shaking equal quantities of water and oil in the presence of surfactant and taking the PIT as the temperature at which separation into a three phase system occurred most rapidly. PIT values quoted are in fact the temperatures at which L3 phase appears in the water/ surfactant systems (52°C and 68°C for $C_{12}E_4$ and $C_{12}E_5$ respectively)[18]. It is likely that the three phase systems which separated were water/L3/triolein, where the detergent phase L3 is not the phase that is associated with PIT and low o/w interfacial tension.

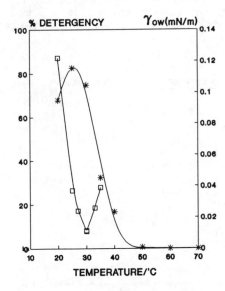

Figure 4. The effect of temperature on γ_{ow} and on detergency for hexadecane in 3×10^{-3} mol dm^{-3} $C_{12}E_4$/10^{-1} mol dm^{-3} NaCl. □, γ_{ow}; *; detergency.

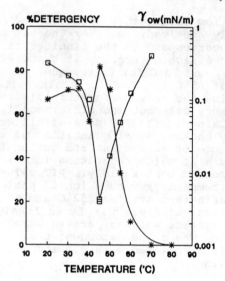

Figure 5. The effect of temperature on γ_{ow} and detergency for hexadecane in 3×10^{-3} mol dm^{-3} $C_{12}E_5/10^{-1}$ mol dm^{-3} NaCl. □, γ_{ow}; *; detergency.

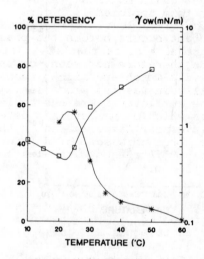

Figure 6. The effect of temperature on γ_{ow} and detergency for triolein in 3×10^{-3} mol dm^{-3} $C_{12}E_4/10^{-1}$ mol dm^{-3} NaCl. □, γ_{ow}; *, detergency.

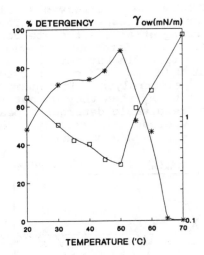

Figure 7. The effect of temperature on γ_{ow} and detergency for triolein in 3×10^{-3} mol dm^{-3} NaCl. $\square$, γ_{ow}; *, detergency.

The Effect of Electrolyte on Anionic Surfactants

Many standard detergent molecules consist of relatively straight chain material which has the general molecular form depicted in Figure 3 where $a_h > a_c$. When the electrolyte level is increased, salting out of the surfactant occurs before the phase inversion structure $a_h = a_c$ is reached. Interfacial tension minima are therefore not seen, or they exist at electrolyte levels which are higher than those encountered normally in washing processes. Essentially, these surfactants are the equivalent of high HLB nonionics which do not exhibit detergency maxima or Phase Inversion Temperatures in the 0-100°C temperature range. For this sort of molecular structure, detergency increases with electrolyte concentration.

Phase inversion conditions can be brought to lower electrolyte levels by increasing the degree of alkyl chain branching so that a_c is larger and a_h can therefore be reduced to equal a_c more easily. Although this is well known in the context of enhanced oil recovery, the principle is not commonly applied in the fabric detergency area.

Figures 8 and 9 show the electrolyte sensitivity of hexadecane and triolein respectively. The maximum in detergency occurring at an electrolyte concentration D^e_{max} and the minimum in interfacial tension correlate quite closely for triolein but not for hexadecane where D^e_{max} is at 0.02 mol dm^{-3} electrolyte and γ_{min} is at 0.035 mol dm^{-3} electrolyte. If indeed there are two mechanisms at work, then they do not both occur or they do not give resolvable detergency peaks in the case of triolein whereas for hexadecane the peak corresponding to γ_{min} does not occur.

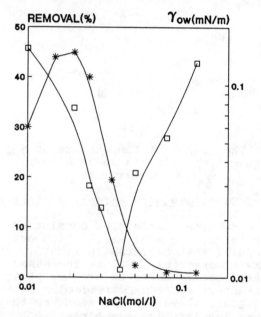

Figure 8. The effect of electrolyte on γ_{ow} and detergency for hexadecane in 3×10^{-3} mol dm^{-3} C_{16} OXS at 25°C. $\square$, γ_{ow}; *, detergency.

The Effect of Composition on Mixtures

The mixing of surfactants to produce synergistic oil detergency is of great importance to the detergent industry and the literature over the last few years reflects this. In 1979 Kubitschek and Scharer[19] reported the use of mixed oil soluble and water soluble nonionics to produce synergistic cleaning of oily soil from hard surfaces. Since then, numerous authors [9,15,24-29] have related oil removal maxima, in systems which include mixtures, to phase inversion conditions, and some have used composition as the primary variable. Among these Schuwuger has examined mixtures of alkyl ether sulphates with linear alkyl benzene sulphonates and also sodium dodecyl sulphate with the nonionic surfactant $C_{16}E_3$. In both cases interfacial tension minima were observed. In the

former case they corresponded approximately with
detergency maxima whereas in the latter case they
appeared at different compositions.

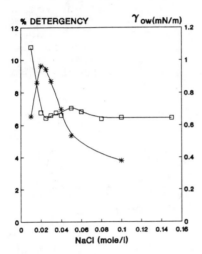

Figure 9. The effect of electrolyte on γ_{ow} and
detergency for triolein in 3×10^{-3} dm^{-3} C_{16} OXS at
25°C. □, γ_{ow}; *, detergency.

Raney, Miller and their co-workers [9,10,20,29] have
also worked with mixtures, both in the sense that they
have used mixed surfactants and also in that they have
shown the effect of surface active impurities
dissolved in the soil on both phase inversion
conditions and the maxima in detergency D^{c}_{max} that they
found to accompany them.

Figures 10 and 11 show the effect of surfactant
composition for SDS/$C_{12}E_3$ mixtures on the cleaning of
triolein and of hexadecane from polyester fabric.
Once again, interfacial tension minima are observed
which have a degree of correspondence with the maxima
in detergency. For hexadecane, the interfacial
tension and oil removal curves are mirror images of
each other. For triolein the agreement is less
dramatic. Indeed there appears to be a slight
mismatch between D^{c}_{max} and γ_{min}. Once again one is
left with a suspicion that the hexadecane removal
curve is the sum of two poorly resolved contributions.
To clarify the situations the experiment was repeated
using the oil squalane, which is known to have a
higher PIT than hexadecane with nonionic surfactants
and would therefore be expected to have γ_{min} at a
higher $C_{12}E_3$ composition in SDS/$C_{12}E_3$ mixtures. Figure
12 shows this to be so, and two detergency maxima are
clearly resolved.

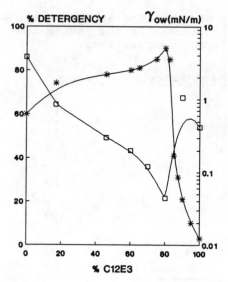

Figure 10. The effect of composition on γ_{ow} and detergency for hexadecane in $SDS/C_{12}E_3$ mixtures at a total concentration of 3×10^{-3} mol dm^{-3} surfactant/10^{-1} mol dm^{-3} NaCl, 25°C. □, γ_{ow}; *, detergency.

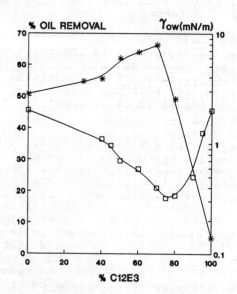

Figure 11. The effect of composition on γ_{ow} and detergency for triolein in $SDS/C_{12}E_3$ mixtures at a total concentration of 3×10^{-1} mol dm^{-3} surfactant/ 10^{-1} mol dm^{-3} NaCl, 25°C. □, γ_{ow}; *, detergency.

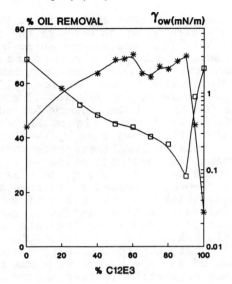

Figure 12. The effect of composition on detergency and γ_{ow} for squalane in SDS/ $C_{12}E_3$ mixtures at a total concentration of 3×10^{-3} mol dm^{-3} surfactant, 10^{-1} mol dm^{-3} NaCl, 25°C. □, γ_{ow}; *, detergency.

Discrimination Between Oil Removal Mechanisms

The data point to the existence of two separate oily soil removal maxima, one of which corresponds to the phase inversion condition (γ_{min}) and one which appears in conditions where, in the terms of Fig.3, $a_h > a_c$. Both maxima do not always appear. The simplest explanation for these observations is that the two peaks are due to different oil removal mechanisms. The two major mechanisms are believed to be roll-up and emulsification and the processes can be distinguished surface chemically using a combination of contact angle and γ_{ow} measurement which permit calculation of work of adhesion (W_a) and work of cohesion (W_c) where,

$$W_a = \gamma_{ow} (1 + \cos \theta)$$

$$W_c = 2 \gamma_{ow}$$

Wa represents the affinity of the oil for the polyester substrate and may be taken as an indicator of the efficiency of the roll-up process (defined as complete drop removal). W_c is indicative of the tendency of an oil drop to fragment and may be taken to reflect the emulsification mechanism (defined as

partial drop removal). This approach has been applied
to the data in Figure 1, which compare the cleaning of
four different oils from polyester fabric by SDS with
that for a 50 mole % SDS/$C_{12}E_3$ mixture, and to the
SDS/$C_{12}E_3$/hexadecane data in Figure 11. Table 1 shows
the γ_{ow} data which refer to the various oil/surfactant
combinations in Figure 1, and Figures 13 and 14 show
W_a and W_c as a function of detergency for SDS and
50/50 SDS/$C_{12}E_3$ respectively. Figure 13, which deals
with the data for SDS, shows that W_a is significantly
lower than W_c. This in itself suggests that roll-up
is more important than emulsification in this system.
Moreover, as would be predicted, W_a decreases with
increasing oil removal, whereas W_c increases when
removal is high, which can only occur if emulsification is irrelevant to this particular removal
process. Figure 14 presents similar data for the
50/50 mole ratio mixture of SDS with $C_{12}E_3$. Here the
sensitivity of detergency to the contact angle (Fig.1)
is less pronounced and the W_a and W_c plots in Figure
14 show that emulsification is less unfavourable. In
fact W_a and W_c are comparable and both decrease with
increasing detergency, implying that both mechanisms
are active.

TABLE 1

γ_{ow} for various oils in the presence of SDS and 50
mole % SDS/$C_{12}E_3$ (3×10^{-3} mol dm^{-3}) both in 0.1 mol dm^{-3}
NaCl

Oil	γ_{ow} (mN/m)	
	SDS	50/50 SDS/$C_{12}E_3$
Hexadecane	3.25	0.25
Triolein	1.05	1.12
Olive Oil	1.31	0.65
Castor Oil	1.05	1.28

Figure 15 shows oil removal and contact angle data for
the SDS/$C_{12}E_3$/hexadecane system. Clearly the increase
in detergency occurs despite a reduction in contact
angle as the $C_{12}E_3$ content increases up to D^c_{max}.
This, at first sight implies that emulsification
becomes more favoured because the decrease in contact
angle operates against roll-up. While this is
certainly true it is noted that complete removal of an
oil drop by a shear field is also directly
proportional to interfacial tension[26], so that a
reduction in γ_{ow} could, in principle, offset the
contact angle decrease. Here however, the contact
angle reduction appears dominant, Figure 16 showing
that, at low levels of $C_{12}E_3$, W_a is very much smaller

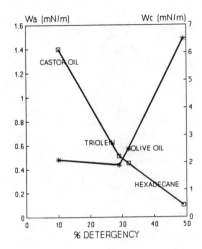

Figure 13. The relationship between detergency, work of adhesion (W_a) and work of cohesion (W_c) for various oils in 3×10^{-3} mol dm^{-3} SDS/10^{-1} mol dm^{-3} NaCl, 25°C. □, W_a; *; W_c.

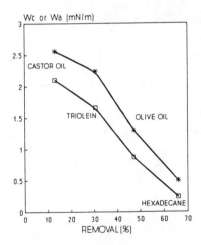

Figure 14. The relationship between detergency, work of adhesion (W_a) and work of cohesion (W_c) for various oils in 1:1 mole ratio mixture of SDS/$C_{12}E_3$ at a total concentration of 3×10^{-1} mol dm^{-3} surfactant, 10^{-1} mol dm^{-3} NaCl, 25°C. □, W_a; *; W_c.

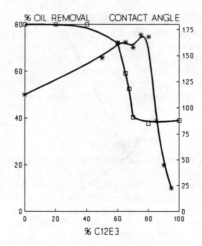

Figure 15. The effect of composition on contact angle and detergency for SDS/$C_{12}E_3$ mixtures at a total concentration of 3×10^{-3} mol dm^{-3} surfactant, 10^{-1} mol dm^{-1} NaCl, 25°C. □, contact angle; * oil removal.

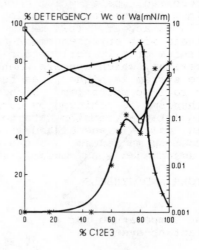

Figure 16. The effect of composition on detergency and on work of adhesion (W_a) and work of cohesion (W_c) for SDS/$C_{12}E_3$ mixtures at a total surfactant concentration of 3×10^{-3} mol dm^{-3}, in the presence of 10^{-1} mol dm^{-3} NaCl at 25°C. □, W_c; *, W_a, +, oil removal.

than W_c. This implies that emulsification is unlikely and that removal takes place through roll-up. As the level of $C_{12}E_3$ increases, W_c falls rapidly until it is of the same order as W_a, which rises to meet it. It is therefore concluded that the increase in detergency as the synergy peak is approached derives mainly from an increase in the efficiency of the emulsification process.

4 SUMMARY

The properties of nonionic surfactants change markedly as a function of temperature, varying from water soluble at low temperature to oil soluble at high temperature. At intermediate temperatures there is a minimum in the oil/water interfacial tension which is associated with phase inversion in emulsions. In contrast the surface activity of anionic surfactants is primarily electrolyte dependent so that minima in o/w interfacial tension can be generated by changes in salinity. For mixtures of surfactants, interfacial tension minima occur when a w/o emulsifier is mixed with an o/w emulsifier. These phenomena are shown to have a common origin in surfactant molecular geometry, and in each case they are loosely associated with a maximum in oily soil detergency for the same surfactant/oil combination. Two types of detergency maximum have been resolved, one of which corresponds exactly to the minimum in interfacial tension, but which sometimes fails to appear. This maximum has been shown to originate with an enhancement of the emulsification mechanism of oily soil detachment. The second detergency maximum, which appears when the effective HLB of the system is higher, has been associated with the roll-up mechanism. Discrimination between the two mechanisms is achieved by combining contact angle and interfacial tension data to estimate the work of adhesion, which is associated with the efficiency of the roll-up mechanism, and the work of cohesion, which is associated with emulsification.

5 ABBREVIATIONS AND SYMBOLS

PIT Phase Inversion Temperature for a Water/Oil/Surfactant system

Wa Work of Adhesion

Wc Work of Cohesion

D^t_{max} Temperature for maximum detergency by a nonionic surfactant

D^e_{max} Electrolyte concentration for maximum detergency by an anionic surfactant

D^c_{max} Composition for maximum detergency by a
surfactant mixture

CPP   Critical Packing Parameter = V/la_h

V   Volume of a single hydrocarbon chain

C   Critical length of a hydrocarbon chain

a_h Head group area

a_c Area of an alkyl chain

γ_{ow} Oil/water interfacial tension

γ_{min} A minimum value of oil/water interfacial tension
for a particular surfactant system

θ Contact angle of an oil drop on a solid substrate
immersed in a surfactant solution. (measured
through the oil phase)

6   REFERENCES

1.   K Durham in "Surface Activity and Detergency"
Ed K Durham. Macmillan, London, 1961, Chapter 1.

2.   M J Schick in Surfactant Science Series, volume
23 "Nonionic Surfactants", Ed M J Schick. Marcel
Dekker, New York 1987, Chapter 13.

3.   R Aveyard and T A Lawless, J. Chem. Soc. Faraday
Trans. 1, 1986 82, 2951.

4.   R Aveyard, B P Binks, S Clark and J Mead,
J. Chem. Soc. Faraday Trans. 1, 1986, 82, 125.

5.   R Aveyard, Chemistry and Industry, 20 July 1987,
474.

6.   R Aveyard, B P Binks, J Mead and J H Clint,
J. Chem. Soc. Faraday Trans. 1, 1988, 84, 675

7.   M J Rosen and D S Murphy, J. Colloid Interface
Sci., 1986, 110, 224.

8.   P M Holland in "Phenomena in Mixed Surfactant
Systems", Ed. J F Scamehorn, ACS Symposium Series
311, 1986, p102.

9.   P M Holland, Colloids and Surfaces, 1986, 19,
171.

10.   B Lindman, S Engstrom and P Backstrom, J Surface
Sci. Technol., 1988, 4, 23.

11. J N Israelachvili, "Intermolecular and Surface Forces", Academic Press, London, 1985.

12. H L Benson, K R Cox and J E Zweig, HAPPI, March 1985, 50.

13. N Azemar, C Solans, J L Parra and J Calbet; Comm. Jorn. Com. Esp., Deterg, 1988, Parr 19, 121.

14. F Schambil and M J Schwuger, Dechema - Monogr., 1989, 114, 333.

15. K Raney and C A Miller, J. Colloid Interface Sci., 1987, 119, 539.

16. K Raney, W Benton and C A Miller, J. Colloid Interface Sci., 1987, 117, 282.

17. F Mori, J C Lim, O G Raney, C M Elsik and C A Miller, Colloids and Surfaces, 1989, 40, 323.

18. D J Mitchell, G J T Tiddy, L Waring, T Bostock and M P McDonald, J Chem. Soc. Faraday Trans. 1, 1983, 79, 975.

19. H F Kubitchek and D H Scharer, Soap/ Cosmetics/Chem Spec., August 1979, 30.

20. F Jost, H Leiter and M J Schwuger, Colloid Polymer Sci., 1988, 266, 554.

21. M J Schwuger, Chapter 1 in "Structure/ Performance Relationships in Surfactants "Ed. M J Rosen, ACS Symposium Series 253, Washington.

22. P Berth and Jeschke; Tenside, 1988, 25, 78.

23. A Stefanescu, S Florescu and M Rob; Revista de Chimie, 1987, 38, (8), 66.

24. H Krussman; Proc. Conf. "Recent Advances in the Detergent Industry" Cambridge 27/28 March 1990.

25. K H Raney, JAOCS., 1991, 68, 525.

26. M Mahe, M Vignes-Alder, A Rousseau, C G Jacquin and P M Adler, J. Colloid Interface Sci., 1986, 126, 314.

Synthesis, Resolution, and Structural Elucidation of Lipidic Amino Acids, Their Homo- and Hetero-oligomers, and Drug Conjugates

Istvan Toth and William A. Gibbons

DEPARTMENT OF PHARMACEUTICAL CHEMISTRY, THE SCHOOL OF PHARMACY, UNIVERSITY OF LONDON, 29–39 BRUNSWICK SQUARE, LONDON, WCIN IAX, UK

INTRODUCTION

The $\propto$-amino acids with long alkyl side chains, the so-called lipidic amino acids (1, X=H, Y=OH, n=7-17, m=1) and their homo-oligomers, the lipidic peptides (1, m>1), represent a class of compounds which combine structural features of lipids with those of amino acids and peptides[1]. One would expect the chimeric nature of these compounds to be reflected in their physical properties; they should be highly lipophilic due to the long alkyl side chains, yet show polar and conformational behaviour characteristic of amino acids and peptides. Interest has been expressed in the potential use of lipidic amino acids as lubricants[2], cosmetics[3], polishes[4] and surface improvers for ceramics[5]. In addition, the lipidic amino acids and peptides could find use as detergents and biocompatible and/or weather-proof coatings. Of particular interest, however, is their possible use as a drug delivery system and drug formulation[6]. The appropriately protected lipidic amino acids and peptides could be covalently conjugated to or incorporated into poorly absorbed peptides and drugs, to enhance the passage of the pharmacologically active compounds across biological membranes.

Preparation of Lipidic Amino Acids

The lipidic amino acids can be prepared by treating the appropriate $\propto$-bromo alkanoic acid with ammonium hydroxide[8-11]. An alternative method of synthesis of amino acids[7] was available, whereby 1-bromoalkanes were treated with diethyl acetamidomalonate in the presence of strong base. Hydrolysis and partial decarboxylation of the alkyl diethyl acetamidomalonate intermediates were effected by heating at reflux in concentrated hydrochloric acid to yield the protonated amino acids.

$$
\begin{array}{c}
CH_3 \\
[CH_2]_n \\
X[-NH-CH-CO-]_mY \qquad \underline{1}
\end{array}
$$

2-amino-decanoic acid was resolved enzymatically using the method of Birnbaum et al.[9], in which the enzyme acylase I was used to specifically hydrolyse the S-*N*-chloroacetyl derivative of the SR-α-amino acid to yield the free S-amino acid. The R-isomer was obtained by hydrolysis of the remaining solution with a mineral acid. However, this method is unsuitable for the longer chain members of the series, due to the inability of the enzyme to hydrolyse the *N*-chloroacetyl derivatives of these compounds. Thus, a chemical method of resolution was used, in which the racemic lipidic amino acid methyl esters were heated at reflux in toluene with the chiral ∝-pinene derivative (1S,2S,5S)-(-)-2-hydroxypinan-3-one[12] in the presence of a catalytic amount of boron trifluoride etherate. The resulting diastereomeric Schiff bases were separated by thin layer chromatography on silica gel. The optically pure methyl esters were obtained by hydroxylamine hydrochloride-assisted hydrolysis of the imines. Saponification of the methyl esters gave the optically active amino acids. The absolute configuration of the methyl esters obtained was determined by circular dichroism (CD) studies. The CD of ester of the enzymatically obtained lipidic amino acid showed at $\lambda=210$nm a positive band for 1S, and a negative one for the 1R compound. Thus, it was possible to assign the absolute configuration of the chemically resolved esters, on the basis of the sign of the Cotton effect of the CD spectra; those exhibiting positive ones were assigned the S-, those with a negative one, the R-configuration.

Synthesis of Partially and Fully Protected Homo-Oligomers of Lipidic Amino Acids

Fully protected homo-oligomers of the lipidic amino acids were prepared in solution phase by coupling appropriate *N*- and *C*-protected species. By stirring the amino acids with *tert*-butyryl dicarbonate in *tert*-butyl alcohol/water at pH 11-12, the *N*-*tert*-butoxycarbonyl (Boc) amino acids [X=(CH₃)₃COCO, Y=OH] were prepared. The amino acid methyl esters (X=H, Y=OCH₃) were obtained as the hydrochloride salts from the amino acids by heating with thionyl chloride in methanol. Reacting the Boc lipidic amino acids with the appropriate lipidic amino acid methyl esters, with the assistance of 3-(3-dimethylaminopropyl)-1-ethylcarbodiimide and 1-hydroxybenzotriazole in dichloromethane gave the fully protected dimers. Removal of the Boc group from the fully protected dimers with 3% hydrochloric acid in methanol gave the *N*-deprotected dimers (m=2, X=H, Y=OCH₃). Addition of a further Boc-protected amino acid to the *N*-deprotected dimer-methylester as described above furnished the fully protected trimers (m=3). By repeating *N*-deprotection and coupling, the tetramer [n=11, m=4, X=(CH₃)₃COCO, Y=OCH₃] was synthesised. The same tetramer can be prepared by coupling the *N*-deprotected dimer with

the *C*-deprotected dimer. *C*-Deprotection can be achieved
by saponification with sodium hydroxide in
water/methanol/chloroform solution. In a similar manner,
the octamer [n=11, m=8, X=(CH₃)₃COCO, Y=OCH₃] was
synthesised from the *C*- and *N*-deprotected tetramers.

Synthesis of Partially and Fully Protected Hetero-Oligomers

The lipidic amino acid oligomers exhibit poor
solubility characteristics with increasing molecular
weight. In order to improve the solubility of the
peptides, to modify their degradation/biodegradation
characteristics and to confer on them cross-linking
potential and further pharmaceutical conjugation
possibilities, several *C*-, and/or *N*-protected hetero-
oligomers were prepared, containing either other amino
acids (**3**) or omega derivatised lipidic amino acids (**4**,
Z=Cl, OH).

$$CH_3$$
$$[CH_2]_7$$
$$(CH_3)_3COCO-HN-CH-CO-CODED\ AMINO\ ACID\ or\ -[NH-CH-CO]_mY$$
$$\underline{3}$$
$$Z$$
$$(CH_2)_n$$
$$\underline{4}$$

The omega substituted α-amino-acids were prepared in
an analogous manner to the unsubstituted lipidic amino
acids from 1,10-dibromodecane and 8-bromooctanol. To
prevent unwanted disubstitution, a two-fold molar excess
of 1,10-dibromodecane was used. Surprisingly, the
hydrochloric acid hydrolysis and decarboxylation of both
diethylacetamidomalonate intermediates resulted in the
omega-chloro α-amino acids. In order to synthesise the
hydroxyl substituted amino acids, it was necessary to
carry out the hydrolysis/decarboxylation in perchloric
acid. The *C*- and *N*-protected omega substituted α-amino
acids were prepared as described for the unsubstituted
analogues.

Synthesis, Activity and Transport of Antiinflammatory Benzoquinolizine-Lipidic Peptide Conjugates

Because of their bifunctional nature, the fatty
amino acids and peptides have the capacity to be
chemically conjugated to drugs with a wide variety of
functional groups. The linkage between drug and lipidic
unit may either be biologically stable (ie. a new drug is
formed) or possess biological or chemical instability
(ie. the conjugate is a pro-drug). In either case, the
resulting conjugates would be expected to possess a high
degree of membrane-like character, which may be
sufficient to facilitate their passage across
membranes[13]. The long alkyl side chains may also have
the additional effect of protecting a labile parent drug
from enzymatic attack.

The benzoquinolizine esters (**5** and **6**, m=0) possess antiinflammatory activity[14)-16)]. The free acid function of the side chain can be conjugated to the amino group of a lipidic carrier forming a relatively stable amide bond. These compounds thus provide a suitable model system to compare the absoption of parent anti-inflammatory drug with its lipidic conjugate.

Synthesis of benzoquinolizine conjugates

A series of lipidic peptide conjugates (**5**, n=7-17, m=1-3) were prepared by reacting the appropriate lipidic amino acid and peptide methyl esters **1**[1)] with the benzoquinolizine acid[14)]. Sodium borohydride reduction of the ester and the conjugates **5** gave a series of alcohols **6** (n=7-17, m=0-3) as the major products. The configuration of the hydroxyl group of compounds **6** is equatorial as anticipated. The radio-labelled alcohols **6** were obtained by reduction of the keto function of **5** with sodium borotritiate.

Ability of benzoquinolizine conjugates to inhibit release of histamin from mast cells

The success of the various lipidic amino acid-drug conjugates is dependent on one of two factors: That either the entire conjugate possesses the required biological activity (ie. that it is a new drug) or that the lipidic peptide delivery sytem is cleaved *in vivo* to yield the active constituent (ie. the conjugate is a prodrug). The series of conjugates of the benzoquinolizine antiinflammatory alkaloid would be expected to fall into the first category of compounds because of the relatively stable amide link between the parent alkaloid and the delivery system. Thus, the anti-inflammatory activity of the conjugates **5**, **6** (m>0) was compared to that of the parent alkaloids (m=0) in order to investigate the effect on biological activity of lipidic peptide conjugation in this instance. The antiinflammatory activity of the conjugates was assessed by their ability to inhibit the release of histamine from

82

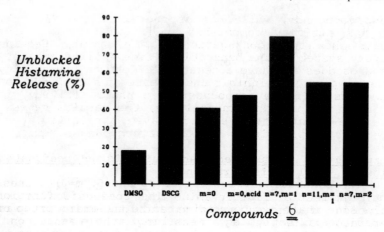

Unblocked Histamine Release (%)

Compounds $\underline{6}$

Fig. 1   Histamine Release

rat mast cells which had been stimulated with Anti-rat
Immunoglobulin E (IgE). Each compound was incubated with
mast cells prior to stimulation of the cells with Anti-
rat IgE. The supernatant obtained after centrifugation
of the cell preparation was assayed for histamine using
a spectrofluorimetric method[17]. Disodium cromoglycate
(DSCG), a known inhibitor of histamine release[18], was
included in each experiment as a positive control. DMSO-
only controls, at the same final concentration as used in
the test compound dilutions, allowed determination of the
effect of DMSO on histamine release. Blanks, containing
cells and all reagents except anti-rat IgE, were included
in each experiment in order to ascertain the extent of
spontaneous histamine release (typically 2-6% of total
histamine release). Total histamine content of the cells
was determined for each experiment by boiling cell
aliquots for 10 minutes and assaying the supernatant for
histamine as described above. Histamine release was
expressed as a percentage of total histamine, calculated
as follows:

$$\frac{\left[\begin{array}{l}\text{Histamine release in the}\\ \text{absence of test compound}\end{array}\right]-\left[\begin{array}{l}\text{Histamine release in the}\\ \text{presence of test compound}\end{array}\right]\times 100\%}{\left[\begin{array}{l}\text{Histamine release in the}\\ \text{absence of test compound}\end{array}\right]}$$

All values were corrected for spontaneous release. The
results are summarised in Fig 1.

The parent benzoquinolizine acid and its methyl ester
show approximately 40% and 50% inhibition of histamine
release from mast cells. The positive control DSCG gives
80% inhibition. The lipidic peptide conjugates inhibit
histamine release to a similar extent to that of the

parent compounds, while one of the conjugate (**5**, n=7, m=7) has activity comparable to that of DSCG. The results show that conjugation of the lipidic peptide delivery system to the benzoquinolizine antiinflammatory compounds does not have a negative effect on the ability of the compounds to inhibit the release of histamine from mast cells. If the compounds exert their activity via the inhibtion of histamine release, the results suggest that the conjugates should have antiinflammatory properties comparable to the parent compounds.

Absorption of ^{3}H-labelled benzoquinolizine conjugates following oral administration

The effect of lipidic amino acid conjugation on the absorption of benzoquinolizine antiinflammatories was determined directly by measuring the amount of radiolabelled compound absorbed following oral administration. The ^{3}H-labelled benzoquinolizine methyl ester **6*** (m=0) and the benzoquinolizine methyl $\propto$-amino-tetradecanoate conjugate **6*** (n=11, m=1), were dissolved and suspended respectively in DMSO, mixed with feed, and fed to starved mice. Blood samples were taken from the animals at regular intervals, and the amount of radioactivity present in the serum determined. Fig 2 summarises the results of these experiments.

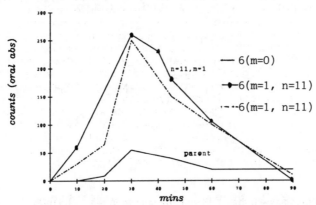

Fig. 2 Oral absorption of the parent benzoquinilizine **6** (m=0) and its lipidic conjugate **6** (n=11, m=1)

The radioactivity detectable in the blood following oral administration of the parent is no greater than background. On the other hand, the conjugate gives peak blood levels approximately five times those of the methyl ester, suggesting that the lipophilic conjugate is absorbed orally (The total radiation detected for the conjugate accounts for only 1% of the total radioactivity administered to the animals).

Penicillin and Cephalosporin Amide Conjugates with Lipidic Amino Acids and their Oligomers

Despite the outstanding clinical success of the β-lactam antibiotics, ineffective absorption of these compounds, particularly following oral administration, has continually plagued investigators in this field. Even compounds that show appreciable activity after oral administration, such as ∝-amino benzyl penicillin (ampicillin) are by no means fully-absorbed from the gastro-intestinal tract[19].

There are several possibilities for conjugation of lipidic amino acids and peptides to the β-lactam antibiotics, by conjugation through either the free carboxylic or the free amino functions. A number of β-lactam antibiotics possess a free amino function which may be acylated with *N*-protected lipidic amino acids, which should provide a convenient way of introducing lipidic functionality into antibiotics.

Synthesis of β-lactam antibiotic amino conjugates[20]

Antibiotics with free amino groups, the 6-amino penicillanic acid (6-APA)[21] and ampicillin (penicillins), and 7-amino cephalosporanic acid (7-ACA)[22] and cephalexin (cephalosporins) were acylated with lipidic amino acids

and peptides **1**. Several procedures for the acylation of amino containing β-lactam antibiotics are known[23]-[27] A mixed anhydride method[28] were used for the preparation of the 6-APA/lipidic amino acid conjugates **7** (n=7-17, m=1-3). Compounds **8** (n=7-17, m=1-3) were prepared using analogous conditions, starting from ampicillin and the Boc-protected lipidic amino acids and peptides. Due to the poor solubility of cephalosporins in aqueous solvents, the cephalosporin derivatives **9** and **10** (n=7-17, m=1-3) were synthesised by an alternative mixed anhydride acylation procedure in organic solvents[29] from either 7-ACA or cephalexin with the appropriate amino acids or peptides.

In vitro antibiotic activity: The minimum inhibitory concentration (MIC) of the compounds was determined *in vitro* against a variety of gram positive and negative bacteria. Most compounds showed moderate to good activity against a non-penicillinase producing strain of *Staphylococcus aureus*. This is interesting from a structure-activity viewpoint, in that lipophilic conjugates of conjugates 6-APA and 7-ACA do not have an aromatic side-chain and very few β-lactam antibiotics with acyclic side chains have been reported. The lipophilic ampicillin conjugate **8** (n=7, m=1) showed activity against a β-lactamase- producing strain of *S. aureus* comparable to that of penicillin G, but significantly weaker than that of ampicillin. The conjugates were active against *Escherichia coli* as well. The ampicillin conjugates **8** were as potent as penicillin

8

G. However, lipidic amino acid conjugation of ampicillin reduced the activity relative to the unconjugated parent compound. The ampicillin conjugates also demonstrated antibiotic activity against a sensitive strain of *Pseudomonas aeruginosa*. All compounds were active against *Clostridium perfringens*. The ampicillin conjugate **8** (n=7, m=1) was more potent than penicillin G and equipotent with cefuroxime.

In vivo antibiotic activity: The conjugates tested *in vivo* were administered by both subcutaneous (s.c.) and oral (p.o.) routes to mice that had been previously infected with a non-β-lactamase producing strain of *S. aureus*. An ED_{50} value was obtained for all conjugates, but only the ampicillin conjugates **8** were active *in vivo* following subcutaneous administration and none of the conjugates were orally active. Thus conjugates **8**, **9** and **10** were not effectively cleaved to the parent antibiotics *in vivo*.

An attempt to couple several neutral and acidic amino acids to phenoxymethylpenicillin and cephalothin has been reported[30], and it was envisaged that the compounds might be better taken up by lysosomes than the parent. However, the conjugates were devoid of antimicrobial activity, presumably due to the stability of the conjugate towards degrading enzymes. The free

carboxylic acid function of the penicillins and
cephalorins is necessary for antibiotic activity and
larger mammals, including man, lack an esterase capable
of hydrolysing simple β-lactam antibiotic esters[31].
Neverthless it should be possible to conjugate lipidic
units to the carboxylic acid, in the hope that enzymatic
hydrolysis will give the active antibiotic *in vivo*. A
degree of biological or chemical instability would
however have to be built into the linking group.

Two types of ester groups suitable for linking
lipidic amino acids and peptides were investigated:
firstly acyloxyalkyl esters which are related to
established enzymically labile pro-drug derivatives[32] and
secondly, novel methoxycarbonyl alkyl esters[36].

Double ester lipophilic derivatives of β-lactam
antibiotics[32]

Double ester derivatives of β-lactam antibiotics
with methylene (**13**, **14**, p=1, m=1-3, n=7-17), ethylene
(**13**, **14**, p=2) and propylene (**13**, **14**, p=3) spacers were
prepared by crown-ether assisted coupling of a
halogenoalkylester of 2-(*tert*-butoxycarbonylamino)
decanoic acid (**12**, p=1-3, m=1-3, n=7-17), to either
penicillin G or cefuroxime.

Double ester prodrugs with a methylene spacer:
Simple esters of the β-lactam antibiotics are known to be
too stable *in vivo* to be used as prodrugs[31], [33]. An
intensive research effort to find a derivative of the
carboxylic acid function of the β-lactam antibiotics that
would improve absorption, yet be sufficiently labile

under biological conditions led to the development of the acyloxymethyl esters of penicillin G[34] and ampicillin[35]. The double ester linkage common to the antibiotic prodrugs, in which the two ester groups are separated by a methylene spacer, is enzymatically-labile. The susceptibility of the methylene-bridged double esters to enzyme-initiated hydrolysis renders them ideal for the bioreversible attachment of lipidic amino acids and peptides to β-lactam antibiotics.

Lipidic amino acids[1] attached to β-lactam antibiotics *via* a double ester of this type would be expected to undergo enzymic hydrolysis to an unstable hydroxy methyl ester intermediate, which would decompose to give the free acid.

$(CH_3)_3COCO\{NH[CH_3(CH_2)_n]CHCO\}_mO(CH_2)_pX$ n=7-17, m=1-3, p=1-3, X=Br, Cl

12

Two pathways to the synthesis of lipidic amino acid

13

double esters of β-lactam antibiotics are conceivable. Either the methylene spacer could be attached to the antibiotic, and the resultant antibiotic alkyl ester esterified to the appropriate *N*-protected lipidic amino acid or peptide, or the alkyl spacer could be added initially to the *N*-protected lipidic amino acid before esterification to the antibiotic. Given the greater chemical lability and cost of the β-lactam antibiotics, it was decided to pursue the second route, in which the antibiotic was not handled until the later stage in the reaction scheme. The chloromethylester **12** (n=7, m=1, p=1, X=Cl) suitable for conjugation to the carboxylic acid group of β-lactam antibiotics was prepared by reacting the potassium salt of the Boc-protected lipidic amino acid **1** with iodochloromethane with the assistance of the macrocyclic ether 18-crown-6. When dibromomethane was used, the desired bromomethylester could not be

obtained.   The major product of the reaction was the
d i e s t e r   **12**   w h e r e   n = 7 ,   m = 1 ,   p = 1 ,
X=OOCCH[(CH$_2$)$_7$CH$_3$]NHCOOC(CH$_3$)$_3$, even when a ten-fold molar
excess of dibromomethane was used.

14

The double ester derivatives **13** where p=1, m=1-3,
n=7-17, X=Boc were prepared by reacting the previously
prepared lipidic esters with the crown-ether complex of
the sodium salt of ampicillin.

The cefuroxime derivatives **14** where p=1, were
synthesised in an identical fashion to that described for
ampicillin conjugates.

Double esters with ethylene and propylene spacers:
Another approach taken to the preparation of labile
esters of the β-lactam antibiotic carboxylic acid group
is exemplified by the methicillin amino ethyl ester
derivative[33], where the ester linkage is believed to be
chemically rather than enzymatically unstable. Similar
behaviour might be expected for analogues **13** and **14** where
p=2 or 3, in which the nitrogen was replaced by an oxygen
atom.

Conjugates were prepared using an ethylene or
propylene bridge between the parent compound and the
lipidic delivery system.   The synthetic approach to
ethylene and propylene lipidic derivatives of β-lactam
antibiotics was identical to that described for the
preparation of methylene analogues. The bromoethylesters
and chloropropylesters **12** where p=2 or 3 were prepared
from the corresponding lipidic α-amino acid and 1,2-
dibromoethane or 1-chloro-3-iodopropane, using the crown-
ether-assisted method.   Double ester ampicillin
conjugates **13** and cefuroxime conjugates **14** were prepared
by reacting the potassium-crown-ether complex of
ampicillin and cefuroxime with halogenoesters **12** where
p=2 or 3.

Double ester conjugates **13** and **14** showed weak or no
antibiotic activity *in vitro*, as expected, but the
compounds were active *in vivo* against a non-penicillinase
producing strain of *S. aureus* following subcutaneous

administration.   These double esters were inactive *in
vitro*, indicating that they underwent hydrolysis *in vivo*
as desired.

Lipophilic methoxycarbonyl alkylesters of ß-lactam antibiotics[36]:

Several series of methoxycarbonyl alkylesters **16-19**
with increased lipophilicity, were prepared by
conjugating penicillin G, ampicillin, cefuroxime and
cephalexin respectively.

$$BrCH[(CH_2)_nCH_3]CO\{NHCH[(CH_2)_nCH_3]CO\}_mOCH_3$$
15

Methoxycarbonyl alkyl esters of the carboxylic acid

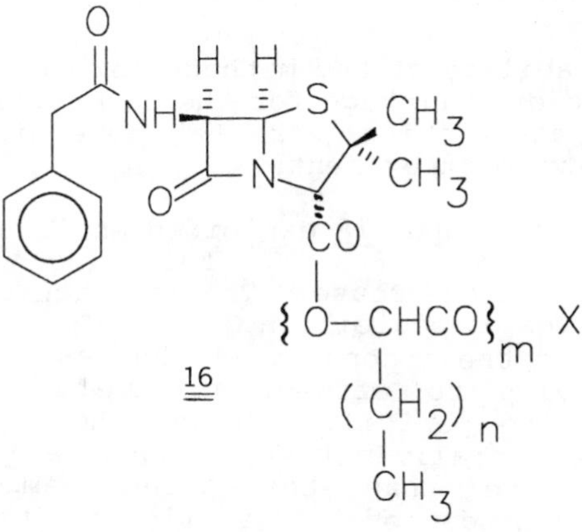

16

group of β-lactam antibiotics are novel derivatives whose
stability in biological systems was not known.   Thus, as
an example of a different type of potentially labile
linkage, lipidic conjugates were prepared from the β-
lactam antibiotics penicillin G, ampicillin, cefuroxime
and cephalexin using 2-bromo alkanoic acid methyl esters
15 (n=7-17, m=0) and the compounds **15** (m=1).   The
monomers were prepared from the corresponding acids with
thionyl chloride in methanol.   The dimeric lipidic
peptide conjugating units were obtained by condensing
methyl 2-aminoalkanoates to 2-bromo-alkanoic acids, using
standard solution phase peptide synthetic methods.   The
lipidic ester conjugates **16** (X=OCH^3) were synthesized by
coupling the bromo-methylesters **15** (m=0) to sodium salt
of penicillin G using a crown-ether assisted coupling
method (Hughes et al., 1991).   Firstly, a complex of the
sodium salt of penicillin G with the macrocyclic ether,
18-crown-6 was prepared, then the complex was reacted
with the 2-bromoalkylmethylesters, furnishing the
ester conjugates **16** in good yield.   This same procedure
was used to synthesise the ampicillin conjugates **17**, the

cefuroxime conjugates **18** and the cephalexin conjugates **19**.

The suitability of the methoxy-carbonyl alkyl ester group as a pro-drug linkage for the carboxylic acid group of β-lactam antibiotics was then determined using *in vitro* and *in vivo* experiments.

Conjugates **16** and **18** exhibited antibiotic activity against *S. aureus* 663 E following subcutaneous administration in the mouse. The most active conjugates were derivatives of penicillin G and cefuroxime. It can be assumed that the secondary alcohol ester linkage was cleaved *in vivo* to afford, presumably, the parent antibiotic. The penicillin G and the the ampicillin conjugates were orally active, conjugate **17** (n=7, m=1) was more active than the parent ampicillin. In conclusion, conjugation of a lipidic moiety *via* a secondary alcohol ester linkage may improve the absorption of β-lactam antibiotics. There appeared to be a preference for short alkyl chains for oral and subcutaneous activity in this series of conjugates, therefore, it can be assumed that the longer alkyl chains protect the ester bond from esterases.

Only one compound, the Penicillin G conjugate **16** (n=7, m=1) showed a weak activity *in vitro* against the sensitive *S. Aureus* strain (663E), *Pseudomonas* and *Clostridia*. The remaining conjugates were inactive.

Lipidic amino acid conjugates with hydrophilic compounds

The underivatised hydrophobic lipidic amino acid-drug conjugates were often insufficiently soluble in water. The problem of how to increase the lipophilicity of the conjugates whilst maintaining adequate water solubility was addressed in two ways. The first was to modify the lipidic amino acids themselves[1], the second was to conjugate the lipidic amino acids with hydrophilic molecules[37]. A number of substituted lipidic acid derivatives, containing hydrophilic groups have been synthesised in our laboratory with the aim of enhancing their solubility in aqueous systems whilst maintaining membrane affinity characteristics. There are several hydrophilic species suitable for conjugation to lipidic amino acids and peptides, such as lactic acid, glycolic acid, glyceric acid and sugars. Of these, sugars represent the most efficient way of increasing water solubility, because of the multiple hydroxyl groups present on one molecule.

Conjugation of lipidic amino acids and oligomers with hydrophilic compounds

Lactic acid was reacted with methyl 2-amino-tetradecanoate using dicyclohexyl-carbodiimide (DCC) as a coupling reagent to yield **20d** dimer. Because of the great difference between the reactivity of the hydroxyl and amino groups under the reaction conditions used, the hydroxyl function of the lactic acid does not need to be protected. Using the same procedure, glycolic acid conjugates **20f**, **20h** and **20i** were synthesised, starting from glycolic acid and esters **20a** - **c** respectively. Base hydrolysis of methylesters **20d** and **20f** resulted in the

free acids **20e** and **20g** with good yields. The coupling of
unprotected D-glucuronic acid with methyl ester **20a**
yielded the new product **20j**. To further increase the
conjugation possibilities and for physico-chemical
investigation, several glycopeptide derivatives **20k** - **o**
were synthesised. The fully-protected **20k** was prepared
by coupling the lipidic amino acid methylester **20c** to the
hydroxyl-protected monosaccharide 1,2:3,4-di-O-
isopropylidene-D-gulonic acid with the assistance of
dicyclohexyl-carbodiimide[38]. The other fully-protected
glycopeptides **20l** and **20o** were prepared in the same
manner, starting from the amino acid methyl ester **20a** and
the N-deprotected lipidic peptide dimer **20b** respectively.
The free acid **20m** was obtained following saponification
of the methyl ester **20l**. The carboxylic acid group of
20m renders it suitable for further coupling to other
compounds. Attempts to remove the two isopropylidene
groups protecting the sugar hydroxyl groups of **20m** were
made with both acetic acid and TFA, each resulting
primarily in the partially-deprotected glycopeptide **20n**.
The second isopropylidene group was found to be resistant
to cleavage even after six hours reflux with acetic acid
or one day with TFA at room temperature.

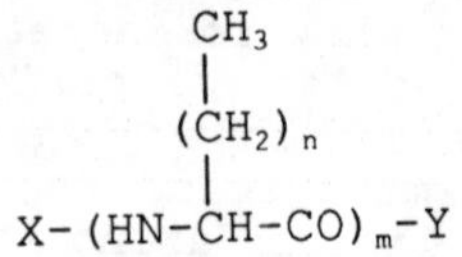

$$CH_3$$
$$|$$
$$(CH_2)_n$$
$$|$$
$$X-(HN-CH-CO)_m-Y$$

20

20	n	m	X	Y
a	11	1	H	OCH_3
b	11	2	H	OCH_3
c	7	1	H	OCH_3
d	11	1	$HOCH(CH_3)CO$	OCH_3
e	11	1	$HOCH(CH_3)CO$	OH
f	11	1	$HOCH_2CO$	OCH_3
g	11	1	$HOCH_2CO$	OH
h	11	2	$HOCH_2CO$	OCH_3
i	7	1	$HOCH_2CO$	OCH_3

$$X_1 =$$

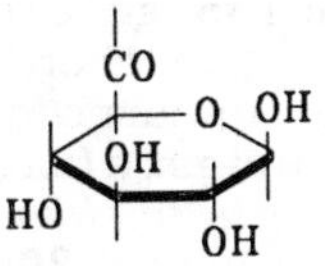

20	**n**	**m**	**X**	**Y**
j	11	1	X_1	OCH_3
k	7	1	X_2	OCH_3
l	11	1	X_2	OCH_3
m	11	1	X_2	OH
n	11	1	X_3	OH
o	11	2	X_2	OCH_3

$$X_2 =$$

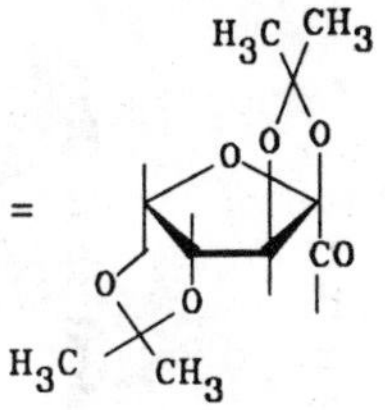

$$X_3 =$$

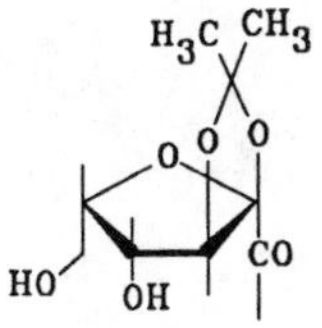

NMR investigations of lipidic amino acid conjugates
Micelle formation in non-aqueous solvents:

The usual picture of the micelle in a non-aqueous solvent is that of the "inverted micelle". The polar head-groups of the surfactant monomer are present in the centre of the micelle with the hydrocarbon chains extending outwards into the solvent[39]. The study of aggregation in organic solvents presents many more problems than are encountered with aqueous solutions. Many of these problems arise because of the smaller aggregates present in organic solvents. As the solutions have little surface activity and the aggregates are not ionized, it is not possible to use the two most commonly used methods to study aqueous systems, i.e. surface tension and conductivity.

It was decided to use ^{1}H-NMR to study the physico-chemical behaviour of lipidic amino acid-systems in organic solvents.

Samples were dissolved in either deutero chloroform ($CDCl_3$) or deutero methanol (CD_3OD), at as high a concentration as possible, as determined by the solubility of the compound, then the solutions were diluted to reach lower concentrations. All spectra were recorded in 0.6 mls of solvent, at 25 °C, as a function of concentration for each of the compounds. Recording the proton NMR spectrum as a function of concentration is an excellent means of monitoring the aggregation of fatty amino acid derivatives. The chemical shift is sensitive to the chemical environment of each proton[40] and can thus be used to observe changes in chemical environment due to liposome/micelle formation, aggregation or dissociation.

For **20d** in methanol (Fig. 3) the chemical shift

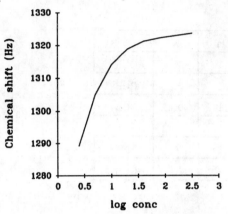

Fig. 3: **Chemical Shift of 2'–H of** 20d

behaviour was monitored over the 320 to 2.50 mM range. The -CH proton from lactic acid was plotted in Fig. 3 (identical results were obtained from other backbone protons). At the highest concentrations the chemical shift of this proton was independent of the concentration, but changed rapidly at lower concentrations. This type of curve was consistent with a fully aggregated form of the molecule at higher concentrations, with a gradual movement towards dissociation of the micelle (or aggregate) at lower concentrations. Even at the lowest concentrations (2.5 mM), the curve did not level off, indicating that a stable plateau, consisting of monomers or the smallest stable forms of the aggregate, was not reached.

The chemical shift of the NH protons of **20f**, **20g** were independent of concentration at low molarities, but began to change rapidly at higher molarities (Fig. 4). At low concentrations, these glycolic acid derivatives were monomers (or very small aggregates) which rapidly aggregated with increasing concentration. In this instance however, the curve did not flatten at the highest concentrations (254 mM and 657 mM respectively), showing that no fully aggregated, stable form was achieved. The additional methyl group present in **20d** appeared therefore to promote aggregation at the expense of the monomer relative to that of the two glycolic acid derivatives.

For monosaccharide derivatives with two and one protecting groups present, **20m** and **20n** respectively, the concentration studies yielded interesting results. Fig. 5 shows the behaviour of the chemical shift of the associated water molecules over the whole of the concentration range studied. The two curves were identical in shape, indicating that both derivatives were exhibiting similar behaviour. The 'S' shaped curve was consistent with the formation of a large, stable aggregate at high concentrations, with a gradual reduction in size with decreasing concentration until monomers, or small aggregates were formed. The approximate critical micelle concentrations (CMC) were determined to be ~16 mM for compound **20n** and ~40 mM for **20m**. The differences between the two CMC values could be explained by considering the polarity of the sugar head groups. The least polar moiety, **20m**, showed a tendency to remain in the monomer form rather than aggregate, whereas the removal of one of these protecting groups, to form **20n**, allowed the possibility of hydrogen-bond formation and electrostatic interaction, via the free hydroxyl groups of the sugar ring. This resulted in micelle (or aggregate) formation at lower CMC values for **20n**.

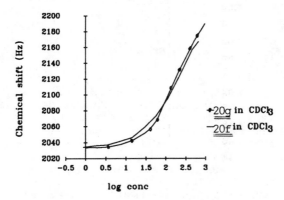

Fig. 4: Chemical Shift of NH proton.

The positions of levelling-off of chemical shift <u>vs</u> concentration plots (Fig. 5) at low concentrations were also different, **20m** forming monomers at lower concentrations than **20n**, again most likely reflecting the increased interaction between molecules via their free - OH groups in **20n**.

The chemical shift parameter however, is merely a reflection of changing chemical environment surrounding an atom, and is not a direct measure of molecular size.

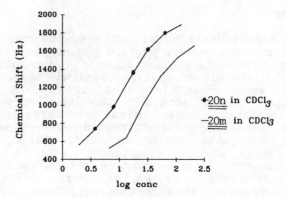

Fig. 5 : Chemical Shift of associated water.

A more reliable measure of this aggregation was to determine the spin-lattice relaxation time[41], or T_1, at each concentration. The resulting graph of T_1 <u>vs</u> concentration should mirror that of the chemical shift <u>vs</u> concentration.

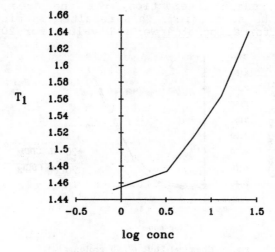

Fig. 6 : Spin Lattice
Relaxation Time (T_1) of 1n.

For derivative **20n**, with one protecting group on the sugar moiety, the plot of T_1 versus concentration (for the -CH$_3$ protons) is shown in Fig. 6. This experiment gave excellent correlation with the shape of that in Fig. 6, proving that the dimensions of the micelles were increasing with increasing concentration and that the observed changes in chemical shift were a reflection of this process.

REFERENCES

1) W.A. Gibbons, R.A. Hughes, M. Charalambous, M. Christodoulou, A. Szeto, A. Aulabaugh, P. Mascagni and I. Toth, Liebigs Ann. Chem., 1990, 1175

2) Kokai Tokkyo Koho (H. Takino, K. Sagawa, N. Kitamura, M. Kilahara Inv.), Jp.Pat. 62/151495 (July 6, 1987) [Chem.Abstr. 107 (1987) P 220318c].

3) Kokai Tokkyo Koho (N. Kitamura, K. Sagawa, M. Takehara Inv.), Jp.Pat 62/4211 (January 10, 1987) [Chem.Abstr. 106 (1987) P 182464s]

4) Kokai Tokkyo Koho (K. Sagawa, M. Takehara Inv.), Jp.Pat. 62/30171 (February 9, 1987) [Chem.Abstr. 107 (1987) P 79672e]

5) Kokai Tokkyo Koho (K. Sagawa, M. Takehara Inv.), Jp.Pat. 62/65964 (March 25, 1987) [Chem.Abstr 107 (1987) P63607b]

6) W.A. Gibbons, Brit.Pat. Appl. GB 2217319 (April 19, 1988).

7) N.F.J. Albertson, J. Am. Chem. Soc. 68 (1946) 450.

8) N. Gerencevic, A. Castek, M. Sateva, J. Pluscec, M. Prostenik, Monatsh. Chem. 97 (1966) 331.

9) S.M. Birnbaum, S.-C.J. Fu, J.P. Greenstein, J. Biol. Chem. 203, (1953) 333.

10) Y. Kimura, Chem. Pharm. Bull. 10, (1962) 1152.

11) Kokai Tokkyo Koho (H. Takino, K. Sagawa, M. Takehara Inv.), Jp.Pat. 63/3078 (January 8, 1988) [Chem.Abstr. 109 (1989) 151601s]

12) T. Oguri, N. Kawai, T. Shioiri, S.-I. Yamada, Chem. Pharm. Bull. 26, (1978) 803.

13) I. Toth, R.A. Hughes, M.R. Munday, C.A. Murphy, P. Mascagni and W.A. Gibbons, Int. J. Pharm., 1991, 68, 191

14) L. Szabo, K. Nogradi, I. Toth, Cs. Szantay, L. Radics, S. Virag and E. Kanyo, Acta Chim. Acad. Sci Hung. Tomus 100: (1979) 19

15) Cs. Szantay, L. Szabo, I. Toth, L. Töke, Gy. Sebes and S. Virag, USA Patent 4,193,998.

16) Cs. Szantay, L. Szabo, I. Toth, S. Virag, E. Kanyo and A. David, U.K. Patent 1,521,320.

17) P.A. Shore, A. Burkhalter and V.H. Cohn, J. Pharmacol. Exp. Ther. 1959, 127, 182.

18) M. Ewnis, A. Truneh, J.R. White and F.L. Pearce, Nature, 1981, 289, 186.

19) W.M.M. Kirby and A.C. Kind, Ann.N.Y.Acad.Sci., 1967, 145, 291.

20) I. Toth, R.A. Hughes, P. Ward, M.A. Baldwin, K. Welham, A.M. McColm, D.M. Cox and W.A. Gibbons, Int.J. Pharm., 1991, 73, 259.

21) F.R. Batchelor, F.P. Doyle, J.H.C. Nayler and G.N. Rolinson, Nature 1959, 183, 257.

22) E.P. Abraham and G.G.F. Newton, Biochem. J., 1961, 79, 377.

23) P. Bamberg, B. Ekström and B. Sjöberg, Acta Chem. Scand., 1967, 21 2210.

24) W.J. Leanza, B.G. Christensen, E.F. Rogers and A.A.

Patchett, Nature, 1965, 207 1395.
25) B. Loder, G.G.F. Newton and E.P. Abraham, Biochem.
    J., 1961, 79, 408.
26) Y.G. Perron, W.F. Minor, C.T. Holdrege, W.J.
    Gottstein, J.C. Godfrey, L.B. Crast, R.B. Babel and
    L.C. Cheney, J. Amer. Chem. Soc., 1960, 82, 3934.
27) C.W. Ryan, R.L. Simon and E.M. van Heyningen, J.
    Med. Chem., 1969, 12, 310.
28) F.P. Doyle, G.R. Fosker, J.H.C. Nayler and H. Smith,
    J. Chem. Soc., 1962, 1440.
29) J.L. Spencer, E.H. Flynn, R.W. Roeske, F.Y. Siu and
    R.R. Chauvette, J. Med. Chem., 1966, 9, 746.
30) Z. Bounkhala, C. Renard, R. Baurain, J. Marchand-
    Brynaert, L. Ghosez and P.M. Tulkens, J. Med. Chem.,
    1988, 31, 976.
31) H.P.K. Agersborg, A. Batchelor, G.W. Cambridge and
    Rule, Br. J. Pharmacol., 1966, 26, 649.
32) R.A. Hughes, I. Toth, P. Ward, A.M. McColm, D.M.
    Cox, G.J. Anderson and W.A. Gibbons, J. Pharm. Sci.,
    1992, August
33) R.P. Holysz and H.E. Stavely, J. Am. Chem. Soc.,
    1950, 72, 4760.
34) H.W. Ferres, Chem. Ind., 1980, 435.
35) W.v. Daehne, E. Frederiksen, W.O. Gundersen, F.
    Lund, P. Morch, H.J. Petersen, K. Roholt, L. Tybring
    and W.O. Godtfredsen, J.Med.Chem., 1970, 13, 607.
36) I. Toth, R.A. Hughes, P. Ward, A.M. McColm, D.M.
    Cox, G.J. Anderson and W.A. Gibbons, Int. J. Pharm.
    1991, 77, 13.
37) I. Toth, G.J. Anderson, R. Hussain, I.P. Wood, E.
    del Olmo Fernandez, P. Ward and W.A. Gibbons,
    Tetrahedron, 1992, 48(5), 923.
38) J.C. Sheehan, G.P. Hess, J. Am. Chem. Soc. 1955, 77,
    1067.
39) D. Attwood, A.T. Florence, "Surfactant Systems"
    Chapman and Hall, London, New York 1983
40) A.E. Derome, "Modern NMR Techniques for Chemistry
    Research" pp 63-76 Pergammon Press, Oxford 1988
41) T.C. Farrar and E.D. Becker, "Pulse and Fourier
    Transform NMR" pp 46-65 Academic Press, New York

Biodegradation of Anionic Surfactants

G. F. White and N. J. Russell
DEPARTMENT OF BIOCHEMISTRY UNIVERSITY OF WALES COLLEGE OF CARDIFF,
P.O. BOX 903, CARDIFF CF1 1ST, UK

1   INTRODUCTION

Fatty acid soaps made by the alkaline saponification of tallow-fat and vegetable-oil-derived acyl lipids have been known and used for thousands of years. The Ancient Egyptians have been credited with the invention of soap, but its use remained very limited until more recent times.[1] In 1842, Edwin Chadwick, a vigorous campaigner for the "Sanitory Movement" in the UK, published his findings on the *Sanitory Condition of the Labouring Population of Great Britain*[2] in which he related the occurrence of high rates of disease and death with filthy living and working conditions. The campaign lobbying was so successful that in 1846 Parliament gave its assent to the *Public Baths and Wash Houses Act* which provided working people with access to baths and soap.[3] The use of soap was promoted further in 1853 when Gladstone repealed the Soap Tax,[4] then running at about 1p per lb of soap, equivalent to about £1 on a bar of soap at today's prices! The sanitary movement spread to Europe and the boom in soap consumption was underway.

Problems with the availability of fats and oils during the periods of war in the first half of this century provided the stimulus for development of synthetic alternatives, particularly in Germany and the UK.[5] Development and diversification of the surfactant sector of the chemical industry after the Second World War not only produced adequate soap substitutes, but also enabled the production of a much wider range of surfactants with properties that were appropriate for applications beyond domestic cleaning, laundry and personal hygiene. Consequently, synthetic surfactants are currently used in a great diversity of domestic, industrial and agricultural applications including laundry and dish-washing detergents, shampoos, toothpastes, proprietary cleaners, degreasers, paints, processed foods, ore-extraction, and pesticide formulations.[6,7] Despite the huge variety of synthetic surfactants commercially available,[8] they may all

conveniently be classified according to the ionic nature
of the hydrophilic group, viz. anionic, cationic,
non-ionic and zwitterionic. Examples of these types are
shown in Figure 1.

The great utility of surfactants is manifest in the
growth in the scale of world-wide commercial production
for all types which is currently 15.9 x 10^6 metric
tonnes per annum.[9] Obviously, the environmental impact
of synthetic surfactants is also associated with this
period but the development of pollution problems has
followed a quite different pattern of temporal change.
Some of the early synthetic anionic surfactants were
highly resistant to biodegradation processes (*vide
infra*), and in the 1950s this led to a number of
anecdotal and documented cases of surfactant pollution
in the environment such as blankets of foam covering

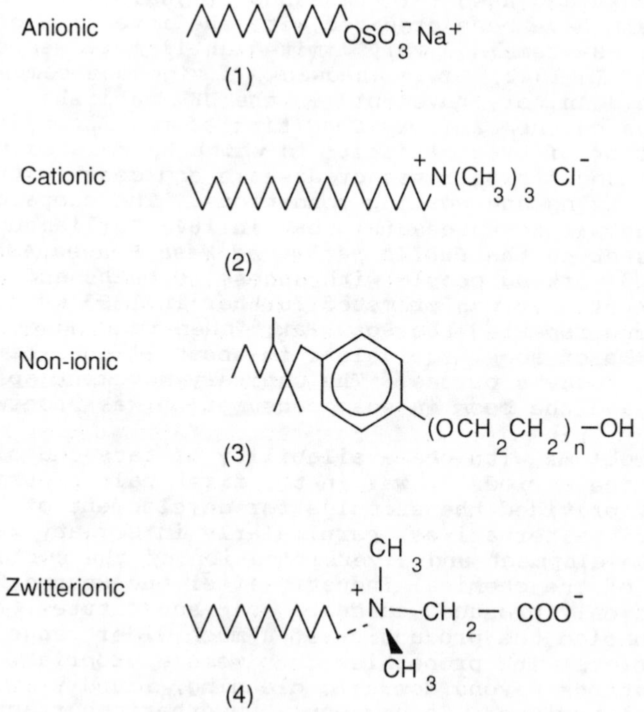

Figure 1 Types of synthetic surfactant classified
according to the nature of the hydrophilic group. The
examples shown are (1) sodium dodecyl sulphate, (2)
octadecyl trimethyl ammonium chloride, (3) nonylphenol
ethoxylate, (4) *N*-dodecyl *NN*-dimethyl glycine.

rivers[1] and water-treatment plants.[10] Such occurrences are now rare in Western countries, because the surfactants in large-scale use are capable of being biodegraded in both engineered (e.g. water-treatment plants) and natural environments which they enter. This has been achieved as the result of extensive research in several countries on the biochemistry of the biodegradation processes occurring in the environment. From these studies, it has been possible to identify those chemical structural features of surfactants which predispose them to rapid biodegradation compared with those features which render the molecule recalcitrant.

Biodegradation of surfactants is accomplished predominantly by aerobic bacteria[10] which in so doing avail themselves of carbon (and also some other elements, depending on the surfactant) present in the surfactant molecule. The anionic surfactants are by some way the most important in terms of scale of production and use. The purpose of this chapter is to present an overview of the biodegradation of surfactants from this perspective of bacterial nutrition, and it focuses on anionic surfactants because of their quantitative importance. Examples of those important anionic surfactants, to which further reference will be made later, are shown in Figure 2.

2   MICROBIAL NUTRITION IN OLIGOTROPHIC HABITATS

It is a fact of microbial life that most microorganisms exist in environments where the nutrient concentration is low.[11] So-called "oligotrophic" environments are often considered among the extremes to which microorganisms have become adapted.[12-14] However, such conditions are so ubiquitous, particularly in the oceans, rivers and lakes on Earth,[11] that paradoxically they are the norm rather than the exception. Faced with such widespread privation, microorganisms have adapted to survive what has been termed a "fast [*sic*] and famine" existence.[15] A central requirement for the survival of microbial cells is that they must be able to generate a minimum metabolic energy in order to sustain their structural and functional integrity. The vast majority of microorganisms are chemo-organotrophs, i.e. they obtain their energy by oxidation of the reduced carbon that is present in organic nutrients. Consequently for many microorganisms, survival depends on scavenging for reduced organic carbon compounds in their environment.

Numerous microbial phenomena have been identified as being possible strategies which microorganisms might use to gain an advantage in nutrient scavenging. These include chemotaxis, the ability to move towards higher concentrations of attractants (nutrients) for example in gradients which exist in stagnant water;[16,17] attachment

to surfaces such as sediments and river-bed stones in
fast-flowing streams;[11,18-20] changes in nutrient
transport systems[21]; and the development of catabolic
diversity to enable the organism to utilize a wider
range of potential nutrients.[21] It is this last point
which brings us back to pollutants such as surfactants.

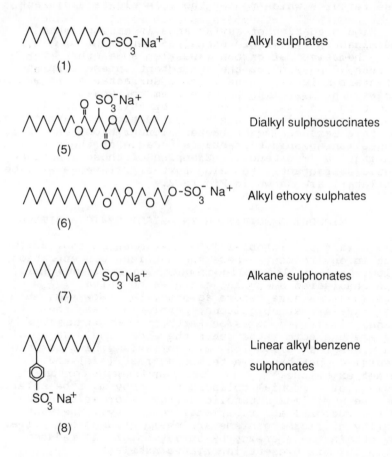

(1)   Alkyl sulphates

(5)   Dialkyl sulphosuccinates

(6)   Alkyl ethoxy sulphates

(7)   Alkane sulphonates

Linear alkyl benzene
sulphonates

(8)

Figure 2 Examples of commercially-important anionic
surfactants. (1) Sodium dodecyl sulphate, (5) sodium
dioctyl sulphosuccinate, (6) sodium dodecyltriethoxy
sulphate, (7) sodium dodecane sulphonate, (8) sodium
3-(*p*-sulphophenyl)dodecane.

3   SURFACTANTS AS MICROBIAL NUTRIENTS

In recent times on the evolutionary timescale, microbial survival has been assisted by human mobilisation of organic carbon in fossil deposits. Hitherto, this organic carbon has been of relatively restricted accessibility to microorganisms because of the localized nature of such deposits. However, the exploitation of fossil carbon for feedstocks in the chemical industry has mobilized this resource into a much wider environment. For example, many synthetic surfactants derive their hydrophobic chains from paraffin fractions in crude oil via intermediates such as olefins and alcohols.[7] Subsequent use of the surfactants leads eventually to their discharge into waters and soils, either directly or via waste-treatment plants, so that the organic carbon previously locked away in mineral deposits now becomes accessible as a potential source of nutrition to microbial populations in the environment.

In addition, agriculture provides tallow and vegetable oils such as coconut oil and palm-kernel oil, the lipids which have been traditional starting materials for the soap industry[8] and are now also used as raw materials for the production of "synthetic" (i.e. non-soap) surfactants.

The key step in production of a surfactant is covalent attachment of the hydrophobic alkyl chains found in petrochemical and agricultural lipid feedstocks, to a hydrophilic group, usually the $-SO_3^-$ group either as a sulphonate or a sulphate ester. This clearly alters the properties of the molecule to human advantage in the intended application, but it also potentially occludes the molecule from the normal catabolic pathways which would otherwise have been involved in its biodegradation. Thus, the biodegradation of surfactants needs to be viewed in terms of how microorganisms exploit their catabolic diversity in response to the challenge of these new organic carbon sources in a nutrient-depleted environment.

Broadly speaking, microorganisms can use two general strategies to gain access to the reduced carbon in the hydrophobic chains of surfactants: either by separating the hydrophilic groups from the hydrophobic group to leave an alkyl chain with a readily-metabolisable functional group such as -OH or -COOH, or by direct attack at the ω-position of the intact molecule with the hydrophilic group still in place. We shall now examine biodegradation pathways of anionic surfactants in relation to these strategies.

4   SEPARATION OF HYDROPHILE FROM HYDROPHOBE

Alkyl Sulphates

Alkyl sulphates produced for use in commercial surfactants contain an ester bond between a long-chain, usually primary, alcohol and sulphuric acid. Biodegradation of these compounds by *Pseudomonas* species[22-24] is initiated by enzymic hydrolysis of the sulphate group to liberate the parent alcohol (9) which then undergoes oxidation via the alkanal (10) to the fatty carboxylic acid (11) (Figure 3). Fatty acids either undergo elongation to C_{16} and C_{18} homologues and incorporation into phospholipids, or they enter central metabolic pathways via β-oxidation to acetyl-Coenzyme A. Secondary alkyl sulphates such as decyl-4-sulphate (12), important in the Teepol surfactants produced by Shell in the early post-war years,[5] follow a similar biodegradation pathway, except that the liberated secondary alcohol, decan-4-ol (13) undergoes C-C bond scission[25] at the substituted carbon atom via a dione intermediate (14). The resulting alkyl-chain-fragments are alkanals and alkanoic acids which then re-enter central metabolic pathways to yield acetyl-CoA.

Alkylsulphatase enzymes catalysing the hydrolysis of alkyl sulphates in *Pseudomonas* spp. have been purified and well-characterized in terms of their substrate specificity and kinetics and the mechanisms of action.[26,27] Bacteria containing alkylsulphatases are commonly found in natural environments.[28-30] Naturally-occurring alkyl sulphates are equally ubiquitous[26]; examples include the sulphated lipids and chlorosulpholipids present in freshwater algae,[31,32] and these compounds may be natural substrates for bacterial alkylsulphatases. Thus, evolution of the biodegradative capacity towards alkyl sulphate surfactants is probably not a recent response to environmental pollution with synthetic surfactants, but dates from much earlier encounters with natural analogues.

Dialkyl Sulphosuccinates

Like the alkyl sulphates, dialkyl sulphosuccinate surfactants (5) also contain ester bonds, but in this case the acid components are the carboxylic acid groups of succinic acid rather than sulphuric acid (Figure 2). Hydrolytic scission of the alcohol from ester linkage is again the mechanism that allows bacteria to gain access to the long-chain alcohol (Figure 4). As with the alkylsulphatases, esterase reactions involving natural analogues[33] can be readily identified in lipid metabolism. In contrast with the alkyl sulphates, the hydrophilic co-product (15) of the primary biodegradation step contains a $-SO_3^-$ unit still in organic linkage with the succinate moiety. This point is considered further below.

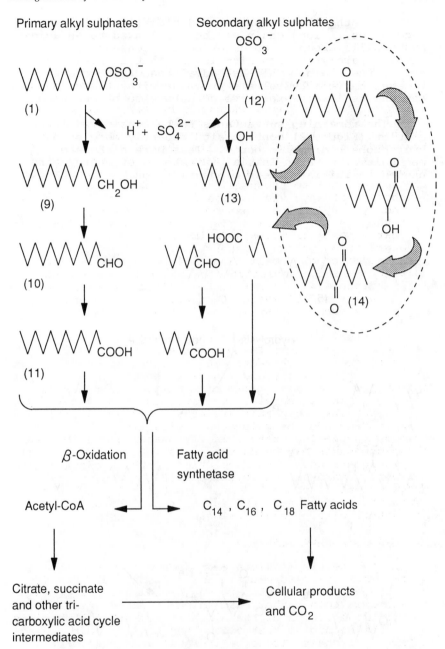

Figure 3 Biodegradation pathways for primary and secondary alkyl sulphates. (1) Sodium dodecyl sulphate, (9) dodecan-1-ol, (10) dodecanal, (11) dodecanoic acid, (12) decyl-4-sulphate, (13) decan-4-ol, (14) decan-4,5-dione. Additional steps required for C-C bond scission of secondary alcohols are shown by hatched arrows.

In both alkyl sulphates and dialkyl sulphosuccinates, a long-chain alcohol is linked to an acidic
hydrophilic group by an ester bond. Consequently a
simple hydrolytic mechanism is all that is needed to
access the chain. This has advantages for bacteria
because enzymic hydrolysis proceeds without the need for
any energetically-expensive cofactors to be consumed.

The remaining surfactants to be considered in this
Section (Figure 2) contain significantly more stable
hydrophobe-hydrophile bonds. Thus more elaborate
mechanisms, with an initial investment of energy, are
needed to initiate their biodegradation.

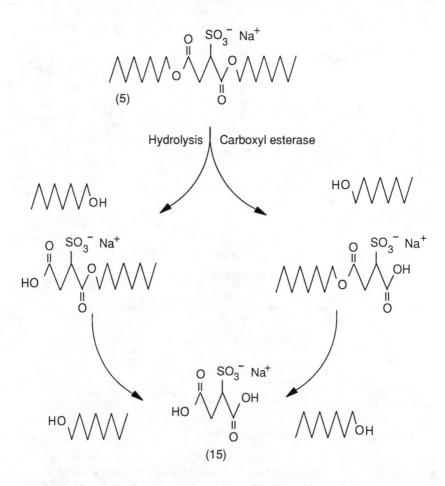

Figure 4 Primary biodegradation pathways for dialkyl
sulphosuccinates. (5) Sodium dioctyl sulphosuccinate,
(15) sodium sulphosuccinate.

Alkyl Ethoxysulphates

Alkyl ethoxysulphate surfactants (6) may be considered as alkyl sulphate analogues in which a poly(ethylene glycol) moiety has been inserted between an alkyl chain and the sulphate group (compare (1) and (6) in Figure 2), and the hydrocarbon chain is attached via an ether bond to poly(ethylene glycol). All *Pseudomonas* spp. so far isolated which can utilize this surfactant for growth, are capable of breaking the alkyl-glycol ether bond in order to gain access to the alkyl chain (Figure 5). Thus, separation of hydrophobic from hydrophilic groups is an important route of primary biodegradation for this type of surfactant. In some bacteria[34] the responsible "etherase" enzyme is very specific for this bond, whereas in other isolates the enzyme(s) are relatively non-specific in so far as they will also attack endo-ether bonds within the poly(ethylene glycol) (Figure 5).

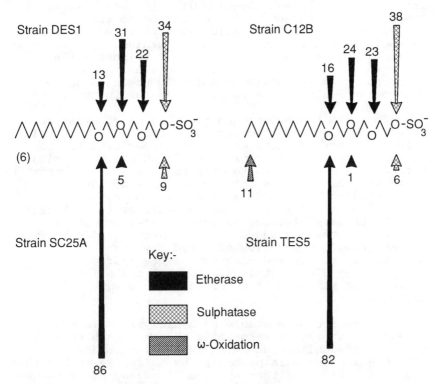

Figure 5 Sites of attack for primary biodegradation of sodium dodecyltriethoxy sulphate (6) by four *Pseudomonas* species employing ether-scission, sulphatase and ω-/β-oxidation mechanisms. Lengths of arrows are proportional to percentages of surfactant (quoted at the foot of each arrow) which undergo primary biodegradation at the indicated point.

Moreover, these surfactants are also substrates for
enzymic hydrolysis of the sulphate group. The alkyl
ethoxylate products of endo-ether cleavage plus sulphate
hydrolysis contain alkyl chains still attached to one or
more hydrophilic ethylene glycol residues. These
compounds therefore serve as substrates in further
rounds of ether cleavage[35] until the alkyl chain is made
available. As will emerge in Section 5, the additional
strategy of ω-/β-oxidation is also operative in some
bacterial isolates, and for completeness this is
included in Figure 5.

Ether-bond scission of a parent surfactant such as
sodium dodecyl triethoxy sulphate, produces metabolites
from the hydrophilic end of the molecule,[36,37] in which
the anionic SO_3^- group remains attached to ethylene
glycol residues (Figure 6). These metabolites
constitute a homologous series of sulphated
poly(ethylene glycol)s containing mono- (16), di- (17)
or tri- (18) ethylene glycol units, and each member of
the series is susceptible to oxidation to the
corresponding carboxylic acid (19)-(21). Not all these
metabolites are formed in single organisms, but in mixed
environmental cultures, all 6 glycol sulphates (16)-(21)
depicted in Figure 6 have been detected.[37] These
compounds clearly retain potentially useful organic
carbon (and sulphur), a point considered in Section 6.

The main hydrophobic co-products of etherase and
sulphatase actions are dodecanol, mono-, di-, and triethylene glycol dodecyl ethers, together with the four
corresponding oxidation products in which the terminal
hydroxyls are converted to carboxylic acids[35] (Figure 6).
There is also evidence for the presence of the
corresponding long-chain alkanals but no clear
indications as to whether the alcohols or the alkanals
are the primary cleavage products. However, other
etherase systems acting on natural ether lipids[38] are
oxidative with production of a hydroxyl on the
hydrophilic moiety and aldehyde on the hydrophobic
group. By analogy, it is likely that the primary
hydrophobic products from surfactant biodegradation are
alkanals.[39]

Ethers are more stable to hydrolysis than are
esters, so it is not surprising that although this is
still a hydrophile/hydrophobe separation strategy, the
etherase mechanism is probably the oxidative rather than
the hydrolytic type that we have discussed hitherto.
Ether lipids are widely distributed[38,40-42] and some, such
as batyl alcohol (3-octadecyloxy-1,2-propandiol), are
structurally similar to these surfactants and are
possible natural substrates for which etherase enzymes
were originally evolved.

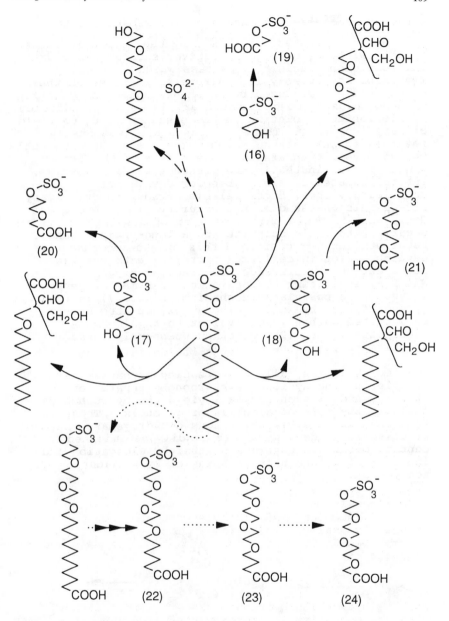

Figure 6 Metabolites produced from sodium dodecyl triethoxy sulphate by sulphatase (- - - -), ether-scission (———) and ω-/β-oxidation (· · · ·). (16)-(18) Mono-, di- and tri-ethylene glycol monosulphates respectively, (19) glycollate-2-sulphate, (20) glycollate-2-(ethoxy sulphate), (21) glycollate-2-(diethoxy sulphate), (22) hexanoic acid-6-(triethoxy sulphate), (23) butyric acid-4-(triethoxy sulphate), (24) glycollate-2-(triethoxy sulphate).

Alkane Sulphonates

Alkyl sulphates are formed by the sulphation of an alcohol with SO_3 or close relatives such as oleum or chlorosulphonic acid,[43] a process which is readily reversible by hydrolysis. Alkane sulphonates on the other hand are produced by sulpho-oxidation of paraffins with SO_2 plus O_2 to give secondary isomers, or addition of bisulphite to olefins in the presence of O_2 to yield primary isomers.[44] These processes are less easily reversed so that alkane sulphonates are much more stable to hydrolysis than are sulphate esters[45]; the latter hydrolyse in minutes in the presence of dilute acids, but sulphonates remain stable for days. Not surprisingly then, bacterial biodegradation of primary alkane sulphonates does not occur by simple hydrolysis. Early work[46] showed that some sort of desulphonation was occurring to yield sulphite and a long-chain alkanal. Later studies[47,48] confirmed this by showing that the desulphonation in cell-free bacterial extracts was dependent on the presence of molecular oxygen and NAD(P)H, indicating the operation of a monooxygenase enzyme. The postulated mechanism (Figure 7) is that the product formed by introducing an oxygen atom at the sulphonated carbon is equivalent to an aldehyde-bisulphite adduct (25) which spontaneously loses sulphite to yield the alkanal.[48]

To summarize, all the foregoing mechanisms are examples of the hydrophile/hydrophobe separation strategy and as such they all yield the hydrophobic chain in the form of alcohols or alkanals. These in turn are oxidized via ubiquitous dehydrogenase enzymes to alkanoic acids[22] which are readily assimilated by central metabolic pathways, either by elongation and incorporation into phospholipids or by β-oxidation to acetyl-CoA.

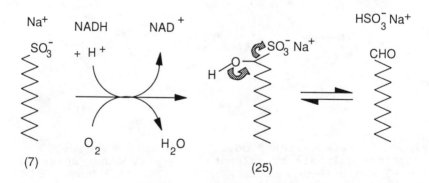

Figure 7 Biodegradation pathway for primary alkane sulphonates. (7) Sodium dodecane sulphonate, (25) sodium 1-hydroxydodecane sulphonate, equivalent to the bisulphite adduct of dodecanal.

5   DIRECT ATTACK ON ALKYL CHAIN

The second strategy for bacterial mobilization of surfactant carbon is direct attack on the alkyl chain in the intact surfactant without removal of the hydrophilic group at the other end of the molecule. In order to biodegrade the alkyl chain by β-oxidation, it is first necessary to introduce a carboxylic acid functional group at the end of the chain. The capacity for terminal-oxidation of alkyl chains is widely distributed in the microbial world, and is important for example in initiating the metabolism of alkanes.[49,50] Insertion of an oxygen atom is achieved by a monooxygenation reaction (Figure 8) analogous to that used to labilize the sulphonate group in alkane sulphonates (Figure 7). The resulting primary hydroxyl group is oxidized subsequently via alkanal to carboxylic acid which then is the starting point for β-oxidation.

At first sight, this seems a more direct route for bacteria to assimilate carbon from the alkyl chain. However, the initiating monooxygenation is NADH-dependent and thus requires an initial investment of metabolic energy. Consequently, the initiation of alkyl-chain biodegradation through ω-/β-oxidation is energetically more demanding than is hydrolytic separation of hydrophile from hydrophobe. Therefore, ω-/β-oxidation is not normally observed where facile separation of hydrophile is possible, e.g. in microbial biodegradation of alkyl sulphates. However, ω-/β-oxidation does occur for those compounds in which separation of hydrophile is more difficult, for example, the alkane sulphonates and alkyl ethoxysulphates

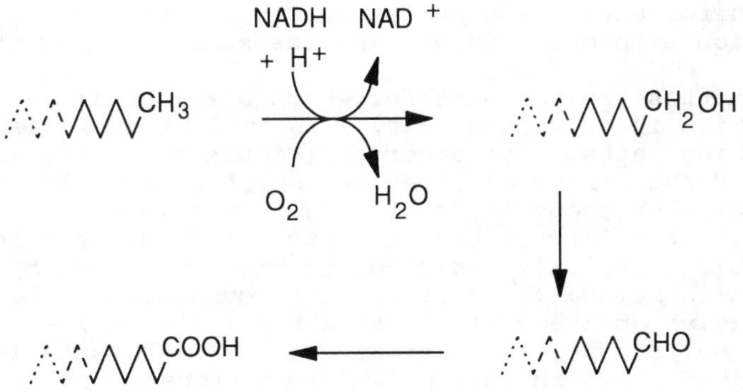

Figure 8 Terminal (ω-) oxidation of alkyl chains

described above. For these surfactants, hydrolytic
scission of the respective C-S or ether bonds is not
easy, and although hydrophile separation does occur, it
is by oxidative, not hydrolytic, mechanisms.
Consequently, besides etherase cleavage, alkyl
ethoxysulphates additionally undergo ω-/β-oxidation
(Figure 5) both in pure[34] and in mixed cultures.[37] The
ω-/β-oxidation of alkyl ethoxysulphates continues until
the site of oxidation approaches quite close to the
hydrophilic end of the molecule (Figure 6).
Intermediates (22)-(24) can be detected which have been
tentatively identified as having the general structure
$^-OSO_2(OCH_2CH_2)_3.O(CH_2CH_2)_n.CH_2COOH$, where n = 0-3 .   Thus,
as with etherase cleavage, primary biodegradation of
this surfactant by direct attack on the alkyl chain
yields organic products which retain potentially
assimilable carbon. This aspect is discussed in Section
6 below.

        Similarly for alkane sulphonates there is evidence
to suggest that ω-/β-oxidation occurs in addition to the
oxidative desulphonation described above.[51]   This is
based on studies of the biodegradation of dodecane
sulphonate in mixed cultures in which disappearance of
the parent surfactant preceded the appearance of any
inorganic sulphate. No direct information is available
to determine how far along the alkyl chain β-oxidation
can progress, but by analogy with the bacterial ω-/β-
oxidation of other surfactants, and with ω-/β-oxidation
of alkane sulphonates in mammalian systems,[52] the process
is unlikely to progress closer than 4-5 carbon atoms
from the sulphonate group.

        Thus for these two groups (alkyl ethoxy sulphates
and alkane sulphonates) in which hydrophile/hydrophobe
separation is more difficult, organisms using the
ω-/β-oxidation pathway can compete effectively. For
some surfactants, hydrophile separation and ω-/β-
oxidation even occur in one and the same organism.[34]

        In those surfactants for which alkyl chain
separation is very difficult, ω-/β-oxidation is the *only*
initiating pathway yet observed (Figure 9). This is the
case for the linear alkylbenzene sulphonates (LAS) such
as 2-(*p*-sulphophenyl)dodecane (26). The inherent
chemical stability of the aromatic carbon-sulphur bond
makes LAS relatively resistant to biodegradation by
hydrophile removal. Under these circumstances, the
alternative strategy of direct attack (i.e. ω-/β-
oxidation) becomes the easiest way by which bacteria can
access the alkyl chain. β-Oxidation procedes until the
site of oxidation approaches the aromatic ring, usually
to within about five alkyl carbon atoms.[53] Further
metabolism of the resulting short-chain sulphophenyl
alkanoates such as 5-(*p*-sulphophenyl) hexanoate (27)
then involves desulphonation and ring cleavage,
usually[54-56] but apparently not always[57] by other

microorganisms constituting a consortium.

The β-oxidative enzymes which are used in surfactant biodegradation seem to be those of the normal fatty acid oxidation pathway[53] which requires an unsubstituted CH_2 group at the β-position and at least one proton on the α-carbon. This pathway can cope with single methyl branches on the α-carbon, but substitution at the β-carbon or *gem*-dimethyl branching anywhere in the chain prevents further progress in oxidation of the alkyl chain. Those alkylbenzene sulphonates produced and marketted in significant quantities during the 1950s were based on tetrapropylene alkylbenzene feedstocks which produced quaternary substitution in the alkyl chains with *gem*-dimethyl branching. Such highly branched alkyl chains are particularly resistant to β-oxidation which led to environmental persistence of these surfactants, and eventual prohibition of their inclusion in domestic detergent formulations.

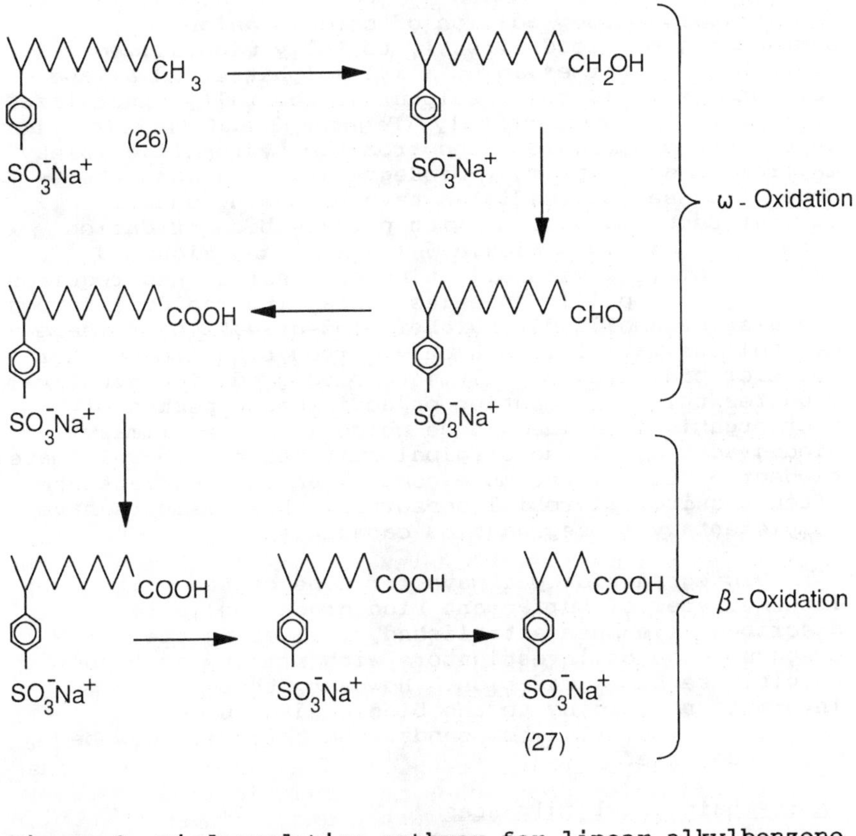

Figure 9 Biodegradation pathway for linear alkylbenzene sulphonates via ω-/β-oxidation.
(26) 2-(*p*-sulphophenyl)dodecane,
(27) 5-(*p*-sulphophenyl) hexanoate.

6   PRIMARY AND ULTIMATE BIODEGRADATION

Separation of hydrophile from hydrophobe (Section 4), and oxidation and shortening of the hydrophobic chain (Section 5) each result in the destruction of the amphiphilic nature of surfactants. Both processes therefore constitute *Primary Biodegradation*, i.e. biodegradation to an extent sufficient to remove some characteristic property of the molecule.[58] Because foam production is an obvious manifestation of surfactant pollution, the achievement of primary biodegradation of surfactants is generally equated with environmental acceptability. Although destruction of surface activity is desirable from a human standpoint, it is even more important to microorganisms because it gives them access to the reduced carbon sources in the alkyl moieties of surfactants. This process goes some way towards *Ultimate Biodegradation* of the surfactant, i.e. biodegradation to CO_2, inorganic salts and "normal" cellular components.

Primary biodegradation of certain anionic surfactants may lead directly to fully mineralized hydrophiles. For example, alkyl sulphates and alkane sulphonates yield their sulphur as the fully mineralized SO_4^{2-} and SO_3^{2-} respectively (Figures 3 and 7). In other instances, organic residues from the hydrophilic moiety may remain after primary biodegradation is complete. Some of these intermediates have been mentioned in earlier Sections dealing with primary biodegradation (see Figure 4 (15), Figure 6 (16-24), and Figure 9 (27)). Complete mineralisation of a surfactant requires that any hydrophilic moieties containing residual carbon are also metabolized. Further biodegradation of these metabolites may yield not only carbon but also sulphur for microbial growth. Ultimate biodegradation usually requires the participation of additional species of microorganisms besides those which initiate primary biodegradation of the original surfactant, i.e. ultimate biodegradation of the more complex anionic surfactants often requires microbial consortia, whose members have complementary biodegradation capacities.

Biodegradation pathways for none of the intermediates of surfactant biodegradation so far described have been established, reflecting the preoccupation of investigators with primary as opposed to ultimate biodegradation. However, there is information relating to the biodegradation of structurally similar compounds, and this will now be described briefly.

Short-chain alkyl sulphates

Combinations of ether-scission and ω-/β-oxidation of alkyl triethoxy sulphates in pure bacterial cultures leads to extensive metabolism of the hydrophobic chain but with simultaneous accumulation of various glycol sulphates (16-18) and oxidized glycol sulphates (19-21) (Figure 6). Although they persist in pure cultures,[34,36] these organic biodegradation intermediates are nevertheless known to be biodegraded by mixed environmental cultures which can complete the conversion of the ester sulphate to inorganic sulphate.[37] There is little direct evidence from which to deduce which metabolic pathways are involved but preliminary studies on the biodegradation of glycollate 2-(diethoxy sulphate) (21) and triethylene glycol sulphate (18) suggest that biodegradation occurs by further cleavage of ether units. Two mechanisms whereby this may occur have been postulated.[59] The main route is considered to be removal of a C_2 moiety from an *oxidized* glycol sulphate to form a glycol sulphate having one fewer glycol unit. The newly-formed terminal hydroxyl of the glycol sulphate is oxidized subsequently by alcohol and aldehyde dehydrogenases to a new oxidized glycol sulphate which then may serve as substrate in another round of C_2 unit removal. This mechanism is analogous to the pathway of poly(ethylene glycol) metabolism proposed by Kawai *et al.*[60-62] The alternative pathway is removal of a C_2 unit from an *unoxidized* glycol sulphate to form a new glycol sulphate with one fewer glycol units, such as proposed for poly(ethylene glycol) metabolism by Pearce and Heydeman.[63]

Inorganic sulphate may be liberated from any of the aforementioned glycol sulphate and oxidized glycol sulphate intermediates, although the corresponding mechanisms and enzymes for these particular compounds have not been elucidated. It is probable that these compounds do not serve as substrates for the alkyl sulphatases responsible for the biodegradation of long-chain alkyl sulphates (Section 4) because these enzymes appear to be specific for unsubstituted alkyl chains with chain lengths longer than five carbon atoms. However, specific mechanisms and corresponding enzymes for the desulphation of some related short-chain alkyl sulphates have been identified.[64]

A coryneform bacterial species, which grows on C_3-C_7 primary alkyl sulphates, gains access to the carbon in these compounds via hydrolysis catalysed by an alkylsulphatase that is specific for C_3-C_7 alkyl chains.[65] In contrast, *Pseudomonas syringae* GG grows well on the secondary alkyl sulphate propyl-2-sulphate but contains no enzyme to hydrolyse this compound. Instead, the ester first undergoes stereospecific oxidation at one of the two terminal methyl groups[66] to give D-lactate-2-sulphate which is the substrate for a stereospecific D-lactate-2-sulphatase.[67] Monomethyl sulphate is reported to undergo sulphatase-mediated hydrolysis in *Hyphomicrobium* sp.[68,69] but there is a monooxygenase-catalysed conversion to formaldehyde and sulphate via methanediol monosulphate[70] in

Agrobacterium sp. M3C. Thus, there is a variety of mechanisms which bacteria use to remove sulphate from short-chain alkyl sulphates. It remains to be seen whether any of these mechanisms is involved in the biodegradation of glycol sulphates, or whether still more systems will be discovered.

Sulphosuccinate

Hydrolytic removal of the long-chain alcohols in dialkyl sulphosuccinates liberates sulphosuccinate (15) itself (Figure 4). This compound is an example of a secondary sulphonate. Although primary sulphonates are common in nature (e.g. taurine, isethionate, Coenzyme M, and the sulphoquinovose moiety in plant sulpholipid), secondary isomers are rare. Consequently, little is known about their biodegradation. Recent studies in our laboratory[71] have shown that a bacterial isolate that grows on sulphosuccinate achieves desulphonation while the 4-carbon chain remains intact, probably as oxaloacetate. This is the first demonstration of desulphonation of a secondary aliphatic sulphonate.

Sulphoacetate and sulphoacetaldehyde

From our earlier discussion of the probable contribution of ω-/β-oxidation to the biodegradation of alkane sulphonates, likely metabolites are short-chain ω-sulphocarboxylic acids such as 4-sulphobutyric acid. The homologous sulphoacetate is known to be metabolized in bacteria participating in the biodegradation of the plant sulpholipid residue, sulphoquinovose.[72] The proposed mechanism involves O_2-dependent and reduced cofactor-dependent cleavage of the C-S bond to yield sulphate and glycollate.[73] In principle a similar mechanism could operate for higher homologues derived from alkane sulphonate surfactants.

The closely-related sulphoacetaldehyde, a deamination product in the bacterial biodegradation of taurine,[74] also undergoes desulphonation in bacteria, but by a quite different mechanism to yield sulphite and acetate.[75] The sulphoacetaldehyde sulpholyase enzyme catalysing the desulphonation is dependent on the presence of thiamine pyrophosphate (TPP) and magnesium ions for its activity.[76,77] In the proposed mechanism, TPP covalently attaches to the substrate by addition at the aldehyde group, from which the original aldehydic proton is removed, leading to β-elimination of sulphite. An enol-keto tautomerization converts the intermediate to an amide which is subsequently hydrolysed to acetate and free TPP. The important aspect of this mechanism, from the point of view of surfactant biodegradation, is the intimate involvement of the aldehydic group in desulphonation. This group is obviously absent from ω-/β-oxidation products of alkane sulphonate biodegradation which probably does not proceed beyond

4-sulphobutyrate. Thus this mechanism is probably not
involved in surfactant biodegradation.

Short-chain sulphophenyl alkanoates

Biodegradation of the alkyl chain in linear
alkylbenzene sulphonates by ω-/β-oxidation yields short-
chain sulphophenyl alkanoates. Ultimate biodegradation
of these intermediates may be viewed as two distinct
processes, namely desulphonation and ring-opening. The
current balance of opinion is that these events occur in
the order given,[78-81] but it must be remembered that this
conclusion is based on studies not with sulphophenyl
alkanoates but with structural analogues. Thus toluene
sulphonic acid is metabolized first by monooxygenation
at the methyl group to produce 4-sulphobenzoate (28)
which in turn undergoes a dioxygenation reaction in
which both oxygen atoms of O_2 are introduced on adjacent
carbon atoms to destabilize the aromatic nucleus (Figure
10). The resulting sulphono-dihydrodiol intermediate
(29) is the equivalent of the bisulphite adduct of an α-
hydroxy-cyclic ketone which is sufficiently unstable

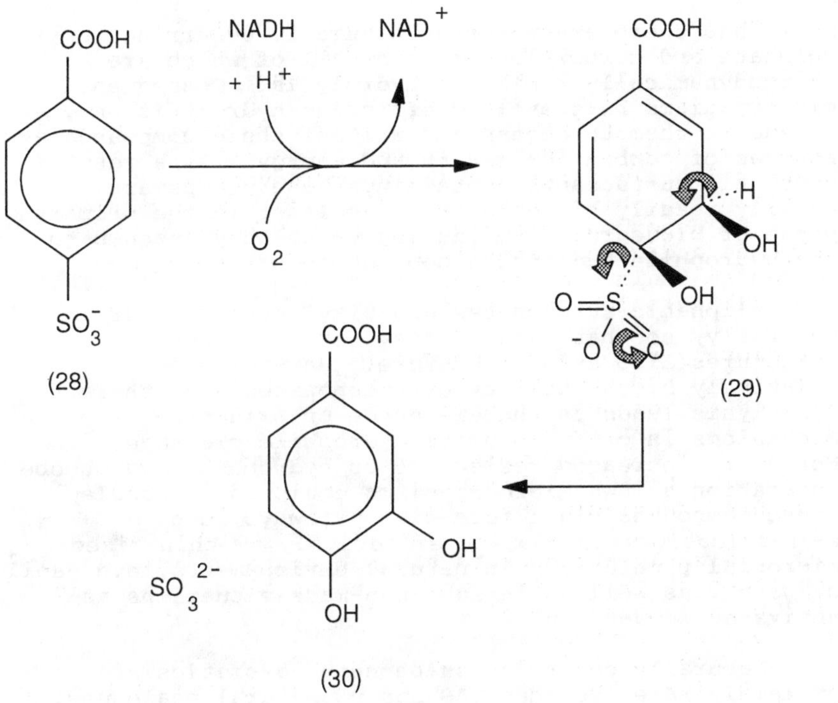

Figure 10 Biodegradation pathway for aromatic
sulphonates. (28) 4-Sulphobenzoate, (29) unstable
sulphono-dihydrodiol intermediate, (30) protocatechuate.

that it breaks down spontaneously with the elimination
of sulphite and an electronic rearrangement to produce
3,4-dihydroxybenzoic acid (protocatechuate). Subsequent
biodegradation of 3,4-dihydroxybenzoic acid via the
so-called *meta*-pathway is well established from studies
on the biodegradation of aromatic compounds in general.[82]
Thus, removal of the highly polar sulphonate group
before ring-opening enables competent organisms to
biodegrade the catechol products using the widely
distributed ring-cleavage enzymes which bacteria have
evolved to assimilate carbon from the numerous aromatic
compounds that occur in nature.

7   CONCLUSIONS

The mechanism by which a given surfactant undergoes
biodegradation is dictated by two factors: first, the
chemistry of the surfactant, in particular the nature of
the linkage between hydrophilic and hydrophobic groups;
and second, the existence in nature of structural
analogues to provide the selective pressure for the
evolution of appropriate catabolic enzymes in
microorganisms.

    Thus there are numerous naturally-occurring
sulphate and carboxylate esters, all of which are
thermodynamically labile to hydrolysis. Consequently,
microrganisms have evolved efficient hydrolytic enzymes
to enable them to access and utilize these compounds as
sources of carbon for growth and energy. As a result,
synthetic surfactants containing these groups are
usually readily biodegradable, at least in the primary
phase of biodegradation, during which they succumb to
the hydrophile/hydrophobe separation strategy.

    Aliphatic sulphonates and alkyl ethers occur
naturally, so that catabolic systems for these
structures also exist in nature. However, the
relatively high stability of sulphonates and ethers to
hydrolysis leads to the emergence of oxidative
mechanisms in order to drive hydrophile cleavage.
Moreover, increased resistance to hydrophile/hydrophobe
separation allows microorganisms equipped to handle
these compounds via alternative pathways, e.g.
ω-/β-oxidation, to compete effectively, within mixed
microbial populations in natural environments, e.g. soil
or water, as well as in such man-made situations as
activated sewage sludge.

    Naturally-occurring sulphonated aromatics are
extremely rare, so that LAS has no natural analogues.
The chemical stability of the aromatic nucleus deflects
initial microbial attack invariably towards
ω-/β-oxidation, with the later stages of desulphonation
and ring cleavage occurring relatively slowly. Indeed,
it is this slow biodegradation of LAS residues which is

currently giving rise to concern about the continued use
of LAS in Europe.

REFERENCES

1. R. B. Cain, in 'Treatment of Industrial Effluents',
    A. G. Callely, C. F. Forster and D. A. Stafford,
    eds, Hodder and Stoughton, London, 1977, p. 283.
2. E. Chadwick, 'Report on the Sanitary Condition of
    the Labouring Population of Great Britain',
    HM Government, Home Department, 1842.
3. Statutory Instrument, <u>Statutes at Large</u>, 1846, <u>18</u>,
    Chap. 260.
4. Statutory Instrument, <u>Statutes at Large</u>, 1853, <u>21</u>,
    Chap. 409.
5. G. F. Longman, 'The Analysis of Detergents and
    Detergent Products', Wiley, Chichester, 1975.
6. D. A. Karsa, 'The Industrial Applications of
    Surfactants', Special Publication 59, Royal Society
    of Chemistry, London, 1987.
7. W. M. Linfield, 'Anionic Surfactants', Surfactant
    Science Series, Vol. 7, parts 1 and 2, Marcel
    Dekker, New York, 1976.
8. G. L. Hollis, 'Surfactants UK', Tergo Data, Durham,
    1979.
9. H. G. Hauthal, <u>Chim. Oggi</u>, 1992, <u>10</u>, 9.
10. R. D. Swisher, 'Surfactant Biodegradation',
    Surfactant Science Series Vol. 18, Marcel Dekker,
    New York, 1987.
11. P. Morgan and C. S. Dow, in 'Microbes in Extreme
    Environments', R. A. Herbert and G. A. Codd, eds,
    Academic Press, London, 1986, p. 187.
12. H. W. Jannasch, in 'Strategies of Microbial Life in
    Extreme Environments', M. Shilo, ed., Verlag Chemie,
    Weinheim, 1979, p. 243.
13. A. L. Koch, in 'Strategies of Microbial Life in
    Extreme Environments', M. Shilo, ed., Verlag Chemie,
    Weinheim, 1979, p. 261.
14. R. Bachofen, <u>Experientia</u>, 1986, <u>42</u>, 1179.
15. J. S. Poindexter, in 'Advances in Microbial
    Ecology', Vol. 5, M. Alexander, ed., Plenum
    Publishing Corporation, New York, 1981, p. 63.
16. J. P. Armitage and J. M. Lackie, eds, 'Biology of
    the Chemotactic Response', Society for General
    Microbiology Symposium 46, 1990, Cambridge
    University Press, Cambridge.
17. G. W. Ordal, <u>Crit. Rev. Microbiol.</u>, 1985, <u>12</u>, 95.
18. J. R. Marchesi, N. J. Russell, G. F. White and W. A.
    House, <u>Appl. Environ. Microbiol.</u>, 1991, <u>57</u>, 2507.
19. M. P. Dawson, B. Humphrey and K. C. Marshall, <u>Curr.
    Microbiol.</u>, 1981, <u>6</u>, 195.
20. K. C. Marshall, <u>Can J. Microbiol.</u>, 1988, <u>34</u>, 503.
21. J. S. Poindexter, in 'Ecology of Microbial
    Communities', Society for Microbiology Symposium 41,
    1987, M. Fletcher, T. R. G. Gray and J. G. Jones,
    eds, Cambridge University Press, Cambridge, p. 283.

22. O. R. T. Thomas and G. F. White, <u>Biotechnol. Appl.</u>
    <u>Biochem.</u>, 1989, <u>11</u>, 318.
23. W. J. Payne and V. E. Feisal, <u>Appl. Microbiol.</u>,
    1963, <u>11</u>, 339.
24. W. J. Payne, <u>Biotechnol. Bioeng.</u>, 1963, <u>5</u>, 355.
25. G. W. M. Lijmbach and E. Brinkhuis, <u>Ant. van Leeuw.</u>,
    1973, <u>39</u>, 415.
26. K. S. Dodgson, G. F. White and J. W. Fitzgerald,
    'Sulfatases of Microbial Origin', CRC Press Inc.,
    Boca Raton, 1982.
27. K. S. Dodgson and G. F. White, <u>Top. Enz. Ferment.</u>
    <u>Biotechnol.</u>, 1983, <u>7</u>, 90.
28. G. F. White, N. J. Russell and M. J. Day, <u>Environ.</u>
    <u>Pollut.</u>, 1985, <u>A37</u>, 1.
29. D. J. Anderson, M. J. Day, N. J. Russell and G. F.
    White, <u>Appl. Environ. Microbiol.</u>, 1988, <u>54</u>, 555.
30. G. F. White, D. J. Anderson, M. J. Day and N. J.
    Russell, <u>Environ. Pollut.</u>, 1989, <u>57</u>, 103.
31. T. H. Haines, <u>Annu. Rev. Microbiol.</u>, 1973, <u>27</u>, 403.
32. E. J. Mercer and C. L. Davies, <u>Phytochem.</u>, 1979, <u>18</u>,
    457.
33. J. L. Harwood and N. J. Russell, 'Lipids in Plants
    and Microbes', George Allen and Unwin, London, 1984.
34. S. G. Hales, G. F. White, K. S. Dodgson and
    G. K. Watson, <u>J. Gen. Microbiol.</u>, 1986, <u>132</u>, 953.
35. E. T. Griffiths, S. G. Hales, N. J. Russell and
    G. F. White, <u>Biotechnol. Appl. Biochem.</u>, 1987, <u>9</u>,
    217.
36. S. G. Hales, K. S. Dodgson, G. F. White, N. Jones
    and G. K. Watson, <u>Appl. Environ. Microbiol.</u>, 1982,
    <u>44</u>, 790.
37. E. T. Griffiths, S. G. Hales, N. J. Russell,
    G. K. Watson and G. F. White, <u>J. Gen. Microbiol.</u>,
    1986, <u>132</u>, 963.
38. F. Snyder, B. Malone and C. Piantadosi, <u>Arch.</u>
    <u>Biochem. Biophys.</u>, 1974, <u>161</u>, 402.
39. E. T. Griffiths, Microbial Degradation of
    Surfactants Containing Ether Bonds, PhD, University
    of Wales, 1985.
40. F. Snyder, 'Ether Lipids: Chemistry and Biology',
    Academic Press, New York, 1972.
41. W. J. Baumann, E. Schupp and J. T. Lin,
    <u>Biochemistry</u>, 1975, <u>14</u>, 841.
42. L. D. Bergelson, V. F. Vaver, N. V. Prokazova,
    A. N. Ushakov and G. A. Popkova, <u>Biochim. Biophys.</u>
    <u>Acta</u>, 1966, <u>116</u>, 511.
43. S. Shore and D. R. Berger, in 'Anionic Surfactants
    Part 1' Surfactant Science Series Vol. 7,
    W. M. Linfield, ed., Marcel Dekker, New York, 1976,
    p. 135.
44. C. Bluestein and B. R. Bluestein, in 'Anionic
    Surfactants Part 2', Surfactant Science Series
    Vol. 7, W. M. Linfield, ed., Marcel Dekker, New
    York, 1976, p. 315.
45. F. C. Wagner and E. E. Reid, <u>J. Am. Chem. Soc.</u>,
    1931, <u>53</u>, 3407.

46. G. Cardini, D. Catelani, C. Sorlini and V. Trecanni, Ann. Microbiol. Enzimol., 1966, 16, 217.
47. G. J. E. Thysse and T. H. Wanders, Ant. van Leeuw., 1972, 38, 53.
48. G. J. E. Thysse and T. H. Wanders, Ant. van Leeuw., 1974, 40, 25.
49. R. J. Watkinson and P. Morgan, Biodegradation, 1990, 1, 79.
50. L. N. Britton, in 'Microbial Degradation of Organic Compounds', D. T. Gibson, ed., Marcel Dekker, New York, 1984, p. 89.
51. T. C. Cordon, E. W. Maurer and A. J. Stirton, J. Am. Oil Chem. Soc., 1970, 47, 203.
52. A. J. Taylor, A. H. Olavesen, J. G. Black and D. Howes, Toxicol. Appl. Pharmacol., 1978, 45, 105.
53. R. B. Cain, Biochem. Soc. (UK) Trans., 1987, 15, 7.
54. L. Jimenez, A. Breen, N. Thomas, T. W. Federle and G. Sayler, Appl. Environ. Microbiol., 1991, 57, 1566.
55. J. C. Sigoillot and M. H. Nguyen, FEMS Microbiol. Ecol., 1990, 73, 59.
56. D. Hrsak, M. Bosnjak and V. Johanides, J. Appl. Bacteriol., 1982, 53, 413.
57. Willetts and R. B. Cain, Biochem. J., 1972, 129, 389.
58. P. A. Gilbert and G. K. Watson, Tenside Deterg., 1977, 14, 171.
59. S. G. Hales, Microbial Degradation of Linear Ethoxylate Sulphates, PhD, University of Wales, 1981.
60. F. Kawai, T. Kimura, T. Fukaya, Y. Tani, K. Ogata, T. Ueno and H. Fukami, Appl. Environ. Microbiol., 1978, 35, 679.
61. F. Kawai, T. Kimura, Y. Tani, H. Yamada and M. Kurachi, Appl. Environ. Microbiol., 1980, 40, 701.
62. F. Kawai, FEMS Microbiol. Lett., 1985, 30, 273.
63. B. A. Pearce and M. T. Heydeman, J. Gen. Microbiol., 1980, 118,
64. G. F. White, K. S. Dodgson, I. Davies, P. J. Matts, J. P. Shapleigh and W. J. Payne, FEMS Microbiol. Lett., 1987, 40, 173.
65. G. F. White and P. J. Matts, Biodegradation, 1992, in press.
66. A. M. V. Crescenzi, K. S. Dodgson, G. F. White and W. J. Payne, J. Gen. Microbiol., 1985, 131, 469.
67. A. M. V. Crescenzi, K. S. Dodgson and G. F. White, Biochem. J., 1984, 223, 487.
68. O. Ghisalba and M. Kuenzi, Experientia, 1983, 39, 1257.
69. O. Ghisalba, H.-P. Schar and G. M. R. Tombo, in 'Enzymes as Catalysts in Organic Synthesis', M. P. Schneider, ed., Reidel Publishing Co., 1986, p. 233.
70. I. Davies, G. F. White and W. J. Payne, Biodegradation, 1990, 1, 229.

122   *Surfactants in Lipid Chemistry*

71. A. Quick, N. J. Russell, G. F. White and S. G. Hales, Biochem. Soc. (UK) Meeting, Swansea, 1991, abst.
72. H. L. Martelli and A. A. Benson, Biochim. Biophys. Acta, 1964, 93, 169.
73. H. L. Martelli and S. M. Souza, Biochim. Biophys. Acta, 1970, 208, 110.
74. H. Kondo, H. Anada, K. Ohsawa and M. Ishimoto, J. Biochem. (Tokyo), 1971, 69, 621.
75. H. Kondo and M. Ishimoto, J. Biochem. (Tokyo), 1972, 72, 487.
76. H. Kondo and M. Ishimoto, J. Biochem. (Tokyo), 1974, 76, 229.
77. H. Kondo and M. Ishimoto, J. Biochem. (Tokyo), 1975, 78, 317.
78. H. H. Locher, T. Leisinger and A. M. Cook, J. Gen. Microbiol., 1989, 135, 1969.
79. H. H. Locher, T. Leisinger and A. M. Cook, Biochem. J., 1991, 274, 833.
80. T. Thurnheer, D. Zurrer, O. Hoglinger, T. Leisinger and A. M. Cook, Biodegradation, 1990, 1, 55.
81. A. M. Cook and T. Leisinger, International Symposium on Environmental Biotechnology, Ostend, 1991, Royal Flemish Society of Engineers, Vol. 1, p. 115.
82. D. J. Hopper, in 'Biodegradation: Natural and Synthetic Materials', W. B. Betts, ed., Springer-Verlag, London, 1991, p. 1.

Degradation of Glycerides by Fungi

J. L. Kinderlerer

FOOD RESEARCH CENTRE, SHEFFIELD HALLAM UNIVERSITY, POND STREET, SHEFFIELD
SI IWB, UK

1. INTRODUCTION

Fatty acids of medium chain length (hexanoic, octanoic, decanoic and dodecanoic C6:0, C8:0, C10:0 and C12:0) are unusual in fats of plant and animal origin.[1] Oils containing these acids are known as the lauric acid oils. The lauric acid oils are derived from the endosperm of two members of the Palmae, the coconut (Cocos nucifera L.)[2] and the oil palm, (Elais guineensis Jacq).[3] Both oils are used in the food and the oleochemical industry.[2-4] World production of the lauric acid oils in 1988 was approximately two million metric tons of which forty percent was imported by countries of the European Community.[5]

Fungi are a major cause of spoilage in stored commodities and probably rank second as a cause of deterioration and loss.[6] Fungi grow over a wide range of environmental conditions (temperature, relative humidity, oxygen concentration and pH).[6] They have the ability to inactivate certain preservatives which may be added to inhibit fungal growth.[7]

Rancidity can occur in refined or edible oils and in processed foods containing edible oils.[8-9] Rancidity is a subjective term which is used to describe the unpalatable odour following oxidative or hydrolytic changes in oils and fats.[9] Traditionally two types of rancidity have been described. The first is hydrolytic rancidity where triglycerides are hydrolysed to give free fatty acids.[10] Free fatty acids can be released from the triglycerides by the activity of lipolytic enzymes (triacylglycerol acyl hydrolases EC 3.1.1.3)[11] as well as by non-enzymatic hydrolysis. The second is oxidative rancidity where mono, di and polyunsaturated fatty acids are oxidised to give hydroperoxides.[9-10] Both the classic free radical mechanism and reactions catalysed by lipoxygenase (linoleate:oxygen oxidoreductase EC 1.13.12) have been implicated.[10] Fat hydroperoxides are odorless and tasteless, but their decomposition products have a great impact on flavour.[8-9,11]

We have described a third type of rancidity, ketone rancidity where medium chain fatty acids are converted into the methyl ketones one carbon atom less than the parent fatty acid by fungal action.[12-15] This type of rancidity has been described as an oxidative variation of the hydrolytic type of rancidity.[14] Relatively little research has been undertaken on ketone rancidity unlike the situation for oxidative and hydrolytic rancidity. Ketone rancidity has been studied in coconut oil,[12,15] palm kernel oil,[15] and desiccated coconut.[14] The relationship between the different types of rancidity is given in figure 1.

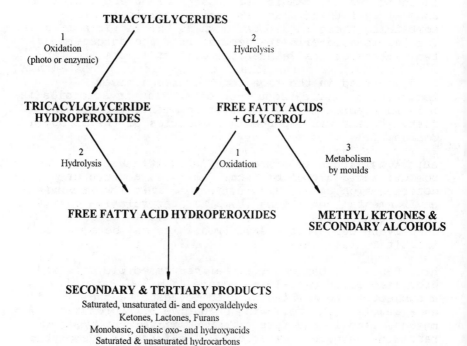

Figure 1: General reaction scheme for (1) oxidative, (2) hydrolytic and (3) ketonic rancidity

Adapted from Robards *et al* 1988[9]

2. TAXONOMIC DISTRIBUTION OF FUNGI CAPABLE OF CONVERTING MEDIUM CHAIN FATTY ACIDS TO METHYL KETONES

The ability of fungi to convert medium chain fatty acids or triglycerides containing these acids has been known for a long time.[12,16] Early work prior to 1948 has been reviewed by Foster in his classic work on the chemical activities of fungi.[16]

Two systematic surveys have been undertaken to determine the distribution of fungi capable of converting free medium chain fatty acids or triglycerides composed of these acids to the corresponding ketone.[17-18] A list of fungi capable of undertaking these bioconversions is given in Table 1. Five genera are important, Aspergillus and Penicillium and their corresponding teleomorphic genera, Trichoderma, Cladosporium and Fusarium. Of the five main divisions of the Fungal Kingdom the ability to bring about these bioconversions has been found in two, the Deuteromycotina and the Ascomycotina. No members of the Basidiomycotina produce methyl ketones although 2-pentanone was isolated from the volatile compounds of Cantharellus cibarius in extremely low yields.[19]

Evidence for these bioconversions in the Zygomycotina is variable. For example, methyl ketones were produced by Syncephalastrum racemosum but not by fifteen other members of the Zygomycotina.[18] On the other hand the German group found this pathway operative in some members of the Zygomycotina (Table 1).[17] The author has isolated and identified 2-heptanone (in low yields) after growth of Mucor javanicus on coconut oil (unpublished work). It would appear that the Mucorales do have a limited ability to bring about these bioconversions.

Fungi with the ability to bring about bioconversion of medium chain fatty acids have been divided into three groups.[18] The first group convert tridecanoin (at pH 7.0) to 2-nonanone in high yields The second again convert tridecanoin to 2-nonanone in high yield as well as converting palm kernel oil to an homologous series of methyl ketones - $(C_{7,9,11,13})$. The last group was defined by the ability to convert tridecanoin to 2-nonanone and an as yet unidentified compound. In this study, only one strain (the Type Culture) was investigated. It is well known that strain differences are important within species. This has been shown for Penicillium caseicolum,[20] P. roquefortii[21] and P. crustosum[7,15,22]. In these studies strains could be grouped with respect to the ability to produce high or low yields of the methyl ketones from the triacylglycerides or free fatty acid.

Table 1
Taxonomic distribution of Fungi capable of converting
dodecanoic acid, palm kernel oil or tridecanoin to
methyl ketones

		Reference:
Zygomycotina	**Zygomycetes**	
	Mucor hiemalis	17
	Mucor plumbeus	17
	Rhizopus japonicus	17
	Rhizopus nigricans	17
	Rhizopus tonkinensis	17
	Syncephalastrum racemosum	18
	(15 tested, 14 negative)	18
Deuteromycotina	**Hyphomycetes**	
	Alternaria kikucuiana	18
	Aspergillus awamori var fumeus	18
	Aspergillus candidus	18
	Aspergillus flavus	17,18
	Aspergillus fumigatus	17
	Aspergillus glaucus	17
	Aspergillus niger	17,18
	Aspergillus oryzae	17,18
	Aspergillus terreus	18
	Aspergillus versicolor	18
	Aspergillus wentii	18
	Cladosporium cladosporiodes	18
	Cladosporium spp	17
	Botrytis cinera	17
	Fusarium avenaceum f. sp. fabae	18
	Fusarium niveum	17
	Fusarium oxysporum f. sp. melonis	18
	Fusarium semitectum	17,18
	Fusarium solani	18
	Gliocladium virens	18
	Gliocladium spp	17
	Neurospora crassa	17
	Paecilomyces farinosus	18
	Penicillium camembertii	17
	Penicillium caseicolum	18
	Penicillium citrinum	18
	Penicillium decumbens	18
	Penicillium glaucum	17
	Penicillium griseo-fulvum	17
	Penicillium roquefortii	17,18
	Penicillium spinulosum	18
	Trichoderma harzianum	18
	Trichoderma hamatum	18
	Trichoderma koningi	17
	Trichoderma paradoxa	18
	Trichoderma polysporum	18

Negative:

Oospora lactis	17
Dematium pullulans	17
Alternaria tenuis	17
Alternaria solani	17
Alternaria gossypina	17
Fusarium sambucinum	17
(50 tested, 27 negative)	18

Ascomycotina

Eurotiales

Chaetosartorya stromatides	18
Dichotomomyces cejpii	18
Emericella nidulans	18
Eupenicillium javanicum	18
Eurotium repens	18
Fennellia flavipes	18
Hamigera striata	18
Hemicarpenteles acanthosporus	18
Neosartoya fischeri var glabra	18

Coronophorales

Gelasinospora tetrasperma	18
Sordaria macrospora	17

Helotiales

Microascus desmosporus	18

Hypocreales

Gibberella lateritium	18
Hypocrea nigricans	18
Nectria flammea	18
Podostroma cordyceps	18

Pezizales

Gymnoascus reessii	18
Monascus auka	18

Dothieales

Mycosphaerella melons	18
Preussia isomera	18
52 species tested, 33 species negative	18

---

3.   TRIACYLGLYCERIDES AS A CARBON SOURCE FOR FUNGAL GROWTH

A major loss of quality of oils such as coconut and palm kernel is due to fungal growth leading to hydrolysis of glycerides and the formation of free fatty acids.[23]   The ability of a fungus to utilise a triacylglyceride as a substrate depends on the fatty acid composition and the slip point. A decrease in the chain length of the

constituent fatty acids will cause a decrease in the slip
point. An increase in the degree of unsaturation of the
constituent fatty acids will again cause a decrease in
the slip point. The composition of a range of oils and
fats is listed in Table 2.

Biomass may be defined as the dry mass of an
organism after growth on a defined medium. It is often
expressed as dry mass cells l^{-1} culture medium. In these
experiments biomass, however, has been expressed as dry
mass cells (mg) substrate g^{-1}. Triacylglycerides provide
a carbon source which is energy rich and carbon poor.
This can be seen in Table 3. Yields of biomass after
growth of P. crustosum on various substrates are
compared. Higher biomass yields were obtained per
kilojoule on sucrose than for any of the triglyceride
substrates. Lower yields were obtained for substrates
containing medium chain fatty acids alone than for
substrates with a mixed medium and long chain fatty acid
composition or just long chain fatty acids, such as
coconut or olive oil.

A similar situation was observed during degradation
of n-alkanes by yeasts; biomass increased with increasing
molecular weight of the substrate from n-nonane up to
n-hexadecane.[24] It is hardly surprising that the relative
growth rates on the two homologous series, n-alkanes and
n-alkanoic acids (fatty acids) are similar as the alkanes
are oxidised to the corresponding fatty acids during
metabolism.[24] The most inhibitory substances were those
containing decanoic acid. Higher yields were found when
the substrate was liquid (compare olive oil with beef
tallow, Table 3).

Table 2

Fatty Acid Composition of Commercial Oils

g per 100 g oil[22]

Results are the arithmetic mean of nine analyses

Fatty Acid	Olive Oil	Beef Tallow	Cocoa Butter	Palm Kernel	Coconut Oil
6:0					0.5
8:0				3.1	7.7
10:0				3.2	5.9
12:0				38.7	46.3
14:0		6.5		14.6	15.6
16:0	10.3	29.0	21.0	7.7	8.3
18:0	2.5	14.9	25.8	2.0	3.1
18:1	62.1	27.4	30.2	12.3	5.7
18:2	8.7	2.7	2.9	2.3	1.8

Table 3

Biomass after Growth on 1 g substrate at pH 7.0 for
72 hours at 25°C and 200 rpm for *Penicillium crustosum*
IMI281919 (inoculum 2.6 x 10^7 spores)[22]

Substrate	Slip Point °C	Energy kJ/g	Yield Dry mass * 100 / mass substrate	mg biomass kJ
Sucrose	–	16.8	18.2	10.8
Liquid triacylglycerides				
Olive oil	-6.0	45.7	17.4	3.8
Trioctadecenoin	5.5	40.2	18.6	4.4
Solid triacylglycerides				
Beef tallow	38.9 – 44.2	43.1	6.6	1.6
Cocoa butter	28.0 – 31.0	43.5	62.0	1.4
Triacylglycerides with medium chain fatty acids				
Coconut oil	23.7 – 26.1	45.7	12.1	2.6
Palm kernel oil	24.5 – 27.3	45.7	3.1	0.7
Trihexanoin	-25.0	29.6	3.6	1.2
Trioctanoin	8.3	32.3	8.7	2.7
Tridecanoin	31.5	34.2	2.0	0.6
Tridodecanoin	46.5	34.8	1.0	0.3
Tritetradecanoin	57.0	36.7	no growth	

4. CONVERSION OF ESTERIFIED MEDIUM CHAIN FATTY ACIDS TO METHYL KETONES ONE CARBON ATOM LESS

Fungal spores,[25-29] vegetative hyphae,[30-33] and cell-free
extracts[28] can convert medium chain fatty acids to the
methyl ketone one carbon atom less. Fungal spores require
additional factors such as glucose, proline, alanine and
serine before this conversion can occur.[28]

Triglycerides containing medium chain fatty acids
can act as substrates for ketone production.[12,15,28] A
ketone one carbon atom less than the parent fatty acid
was produced by fermentation of simple trilglycerides
with vegetative hyphae (Table 4). Thus 2-pentanone was
produced from trihexanoin, 2-heptanone form trioctanoin,
2-nonanone from tridecanoin and 2-undecanone from
tridocenoin.

Table 4

Yield of methyl ketones from fermentation of 1 g simple
and mixed triacylglycerides for 72 hours at pH 7.0 by
Penicillium crustosum IMI 281918.[15,34]

Bioconversion:	Substrate μmol g⁻¹	Product μmol g⁻¹	Yield %
Hexanoate to 2-pentanone			
Trihexanoin			
20°C	7756	489	6.3
25°C	7756	531	6.8
30°C	7756	38	0.5
Coconut oil	Trace amounts of 2-pentanone		
Palm kernel	No 2-pentanone detected		
Octanoate to 2-heptanone			
Trioctanoin			
20°C	6256	468	7.5
25°C	6256	322	5.1
30°C	6256	86	1.4
Coconut oil			
20°C	535	17	3.1
25°C	535	105	19.6
30°C	535	56	10.5
Palm Kernel			
20°C	216	17	3.5
25°C	216	14.3	6.6
30°C	216	14.0	6.5
Decanoate to 2-nonanone			
Tridecanoin			
20°C	5105	3.1	<0.1
25°C	5105	431	8.4
30°C	5105	31	0.6
Coconut oil			
20°C	344	23	6.6
25°C	344	110	32.0
30°C	344	69	20.0
Palm Kernel			
20°C	186	11.8	6.3
25°C	186	21.5	11.6
30°C	186	25.1	13.5
Dodecanoate to 2-undecanone			
Tridodecanoin			
20°C	4473	5	0.1
25°C	4473	22	0.5
30°C	4473	35	0.8
Coconut oil			
20°C	2309	63	2.7
25°C	2309	572	24.8
30°C	2309	191	8.3
Palm Kernel			
20°C	1930	72.2	3.7
25°C	1930	191.5	9.9
30°C	1930	194.3	10.1

An homologous series of methyl ketones $C_{5,7,9,11\ \&\ 13}$ were produced on fermentation of coconut and palm kernel oil. Coconut oil was a better substrate for the formation of ketones than palm kernel oil due to the higher concentration of medium chain fatty acids (Table 2). Higher conversion rates were seen in mixed triglycerides than in simple triglycerides probably due to the presence of long chain fatty acids in the former. Decanoic acid was converted to 2-nonanone at higher rates than the other medium chain fatty acids in both simple and mixed triglycerides. More conversion of acid to ketone occurred in liquid simple triglycerides. (tri hexanoin, trioctanoin and tridecanoin) than in solid ones (tridodecanoin). However considerable conversion of dodecanoic acid occurred in the mixed triglyceride (coconut and palm kernel) but relatively little conversion occurs on fermentation of tridodecanoin. 2-Undecanone was the major product of fermentation of coconut and palm kernel oil reflecting the high concentration of this acid compared to the other medium chain fatty acids in the substrates. The rates of conversion of medium chain fatty acids to methyl ketones in simple triglycerides were C10:0>C6:0>C8:0>C12:0 and in mixed triglycerides were C10:0>C12:0>C8:0 at 25°C.

5. CONTROL OF KETONIC RANCIDITY

Most preservation factors operate through the inhibition or slowing of microbial growth. Both extrinsic factors related to the environment and intrinsic factors related to the food are used.[35] Modern methodology would favour using a number of extrinsic and intrinsic factors together rather than one alone.[36]

In a study on methods of controlling ketonic rancidity,[15,22] it was found that no growth or ketone formation took place at less than 4°C or greater than 37°C after 72 h growth of P. crustosum on coconut oil. Neither did growth or ketone production take place at a water activity (a_w) of less than 0.91 where NaCl (15% m/v) or erythritol (40% m/v) was used to control water activity (a_w). Ketone production was not observed in the absence of oxygen. In all experiments the final pH tended to pH 6.5 - 7.0 irrespective of the initial pH,[34] and the pH did not appear to be useful in controlling ketone production over a 72 hour time period due to the ability of the fungus to alter the environment. Two preservatives, sorbic acid (1000 mg/kg) and Natamycin (1µg/kg) inhibited ketone production although sorbic acid (2,4-hexadienoic acid) was converted to pentadiene[7]. The most obvious control method, however, was to ensure the absence of all fungal spores - asepsis.

6.   PATHWAY FOR THE CONVERSION OF TRIACYLGLYCERIDES
     CONTAINING MEDIUM CHAIN FATTY ACIDS TO THE
     METHYL KETONE ONE CARBON ATOM LESS

The first step in the metabolism of triacylglycerides is
hydrolysis where the fatty acid esterified to the primary
hydroxyl group of glycerol is released.[23] Extracellular
lipases are widespread in the Fungal Kingdom.[37] Fatty
acids with a chain length of less than twelve carbon
atoms are cleaved more readily than the normal chain
length fatty acids.[23] Uptake of the free fatty acids by
spores or vegetative mycelia depends on the pK_a as only
the undissociated form is taken up by the cells.[27-28,30-32]

The main pathway for fatty acid oxidation involves
β-oxidation with the liberation of two-carbon acetyl CoA
molecules.[23] In eukaryotic organisms β-oxidation can
occur at two sites, the mitochondrion and the microbody[34]
(glyoxosome/peroxisome). Certainly both organelles are
found in filamentous fungi and yeasts.[38] When β-oxidation
takes place in the microbody hydrogen peroxide rather
than ATP is formed.

It is accepted that the conversion of medium chain
fatty acids to the methyl ketone involves a partial
β-oxidation.[39] Methyl ketone formation involves two
hydrolytic steps, the initial hydrolysis of triacylglycerides as well as deacylation of the β-oxoacyl-CoA
ester to give the corresponding β-oxoacyl-acid.[28] In
Neurospora crassa a multifunctional protein composed of 3
enzymes (2-enoyl-CoA hydratase, L-3-hydroxyacyl-CoA
dehydrogenase and 3-hydroxyacyl-CoA epimerase) was
isolated from a microbody.[40,41] These authors felt that
β-oxidation of fatty acids took place in the microbody as
only one enzyme having just 2-enoyl-CoA hydratase
activity was isolated from the fungal mitochondrion. One
could speculate that β-oxoacylCoA deacylase is found in
the microbody of certain fungi. It is the presence of two
enzymes, β-oxoacyl-CoA deacylase and β-oxoacyl acid
decarboxylase[42,43] which leads to the production of methyl
ketones. Both these enzymes have been isolated from some
filamentous fungi. Once formed, methyl ketones appear to
accumulate in the culture medium.[18] When oxygen is
limiting the methyl ketone can undergo reduction to give
the secondary alcohol.[13] This reduction is chiral and
may well be of value in the production of fragrances.[44]

Two hypotheses have been put forward to explain why
some filamentous fungi produce methyl ketones. Stokoe[12]
suggested that the conversion of medium chain fatty acids
to the ketone one carbon atom less was a detoxification.
He demonstrated that medium chain fatty acids, C6:0,
C8:0, C10:0 and C12:0 inhibited fungal growth. We[34] have

come to a similar conclusion as Stokoe 63 years later.[12] Lawrence, on the other hand, suggested that the partial β-oxidation allowed a rapid re-cycling of Coenzyme A in an environment where this factor may be limiting.[28]

In conclusion, triacylglycerides containing medium chain fatty acids can undergo an unusual type of fungal degradation whereby volatile methyl ketones one carbon atom less are produced. The odour of the homologous series of methyl ketones changes from pleasant for 2-pentanone (pear-drops) to rancid almond for 2-heptanone, to turpentine-like for 2-nonanone and 2-undecanone.[14] Undoubtedly, the odour or off-odour will depend on the concentration of these compounds. Methyl ketones are used as food flavours and as fragrances in perfumery.[18, 44] It is the presence of the ketones rather than the free fatty acids which are responsible for rancidity of saturated oils of the coconut and palm kernel type.

REFERENCES

1. J. L. Kinderlerer, Int. Biodet., 1986, 22S, 41-47.
2. F. V. K. Young, J.A.O.C., 1983, 60, 374.
3. J. A. Cornelius, Prog. Chem. Fats Other Lipids, 1977, 15, 5.
4. K. G. Berger and S. H. Ong, Oleagineux, 1985, 40, 122.
5. Stad. Commodity Yearbook, 1990, 188 and 192.
6. J. I. Pitt and A. D. Hocking, 'Fungi and Food Spoilage', Academic Press, Sydney, 1985.
7. J. L. Kinderlerer and P. V. Hatton, Food Addit. Contam., 1990, 7, 657.
8. D. A. Forss, Prog. Chem. Fats Other Lipids, 1973, 13, 177.
9. K. Robards, A. F. Kerr and E. Patsalides, Analyst, 1988, 113, 213.
10. R. J. Hamilton, 'Chemistry of Rancidity in Foods', Leatherhead Symposium Proceedings, 1990, No. 47.
11. E. N. Frankel, Prog. Lipid Res., 1982, 22, 1.
12. W. M. Stokoe, Biochem. J., 1928, 22, 80.
13. J. L. Kinderlerer and B. Kellard, Phytochem., 1984, 23, 2847.
14. B. Kellard, D. M. Busfield and J. L. Kinderlerer, J. Sci. Food Agric., 1985, 36, 415.
15. J. L. Kinderlerer and P. V. Hatton, J. Appl. Bact., 1991, 70, 502.
16. J. W. Foster, 'Chemical Activities of Fungi', Academic Press, New York, 1949.
17. W. Franke and W. Heinen. Archiv für Mikrobiol., 1958, 31, 359.
18. T. Yagi, M. Kawaguchi, T. Hatano, F. Fukui and S. Fukui, J. Ferment. Bioengin., 1990, 70, 94.

19. H. Pyysalo, Acta Chem. Scand., 1976, B30, 235.
20. G. Lamberet, B. Auberger, C. Canteri and J. Lenoir, Rev. Lait. Francais, 1982, 406, 13.
21. G. Larroche, B. Tallu and J-B. Gross, J. Ind. Microbiol., 1988, 3, 1.
22. P. V. Hatton, Ph.D. Thesis, CNAA, 1990.
23. M. I. Gurr and J. L. Harwood, 'Lipid Biochemistry', Chapman and Hall, London, 1991.
24. C. A. Boulton and C. Ratledge, 'Topics in Enzyme and Fermentation Biotechnology', Ellis Horwood, Chichester, 1984, p11.
25. R. F. Gehrig and S. G. Knight, Nature, 1961, 192, 1185.
26. R. F. Gehrig and S. G. Knight, Appl. Microbiol., 1963, 11, 66.
27. R. C. Lawrence, J. Gen. Microbiol., 1966, 44, 393.
28. R. C. Lawrence, J. Gen. Microbiol., 1967, 46, 65.
29. C. K. Dartey and J. E. Kinsella, J. Agric. Food Chem., 1973, 21, 933.
30. R. C. Lawrence and J. C. Hawke, J. Gen. Microbiol., 1968, 51, 289.
31. H. L. Lewis, J. Gen. Microbiol., 1971, 63, 203.
32. H. L. Lewis and D. W. Darnall, J. Bacteriol., 1970, 101, 65.
33. T. Yagi, M. Kawagachi, T. Hatano, F. Fukui and S. Fukui, J. Ferment., 1989, 68, 188.
34. P. V. Hatton and J. L. Kinderlerer, J. Appl. Bact., 1991, 70, 401.
35. D. A. A. Mossel, 'Food Microbiology', eds T. A. Roberts and F. A. Skinner, Academic Press, London, 1983, p1.
36. G. W. Gould, M. H. Brown and B. C. Fletcher, 'Food Microbiology' eds T. A. Roberts and F. A. Skinner, Academic Press, London, 1983, p68.
37. S. Oi, A. Sawada and Y. Satomora, Agric. Biol. Chem., 1967, 31, 1357.
38. N. E. Tolbert, Ann. Rev. Biochem., 1981, 50, 133.
39. F. W. Forney and A. J. Markovetz, J. Lipid Res., 1971, 383.
40. R. Thieringer and W-H. Kunau, J. Biol. Chem., 1991, 266, 13110.
41. R. Thieringer and W-H. Kunau, J. Biol. Chem., 1991, 266, 13118.
42. D. W. Hwang, Y. J. Lee and J. K. Kinsella, Int J. Biochem., 1976, 7, 165.
43. W. Franke, A. Platzeck and G. Eichhorn, Archiv für Mikrobiol., 1961, 40, 73.
44. E-H. Engel, J. Heidles, W. Albrecht and R. Tressl, American Chem. Soc. Symposium Series, 1989, 388, 8.

Mycobacterial Phenolic Glycolipids

S. Hartmann and D. E. Minnikin

DEPARTMENT OF CHEMISTRY, THE UNIVERSITY, NEWCASTLE UPON TYNE NEI 7RU, UK

1 INTRODUCTION

The history of phenolic glycolipids began with a study
carried out by Smith *et al.* examining the lipid
extracts of bovine and human tubercle bacilli by infra
red spectroscopy.[1] His findings indicated an aromatic
substituent in the, then named, glycolipids G. These
aromatic glycolipids were later detected in
Mycobacterium kansasii and *Mycobacterium bovis* and were
called G_A and G_B. These characteristic glycolipids were
later called mycoside A and mycoside B.[2] Three 6-
deoxyhexoses (2-*O*-methyl-fucose, 2-*O*-methyl-rhamnose,
2,4-*O*-dimethyl-rhamnose) were detected in these
lipids.[3-5] The structure of the lipid core of mycoside
A and B was partially elucidated by Smith[6] and Noll.[7]
The full structural elucidation of the phenol
phthiocerol was achieved later with the help of [1]H-
nuclear magnetic resonance (NMR) and electron impact
mass spectrometry (EI-MS) of the deacylated lipid core
and fragments from an oxidative degradation.[8,9]

These glycosyl phenolphthiocerols or phenolic
glycolipids are found in only a limited number of
mycobacterial species, some only in certain strains.
Interest in these lipids has been reactivated after
their antigenic properties had been established. There
is a close structural similarity between the phenolic
glycolipids and phthiocerol dimycocerosates, abundant
waxes in the same mycobacterial species (Figure 1).
Both contain a long carbon chain with a β-diol unit and
an α-methyl branched methoxy unit. The major component,
phthiocerol A, is usually accompanied by lesser amounts
of waxes based on phthiocerol B, phthiodiolone A and
phthiotriol A (Figure 1). Phenolphthiocerols carry a
para-substituted phenol ring at the end of the carbon
chain, which is linked to an oligosaccharide containing

Phthiocerol dimycocerosates

Phthiocerol A

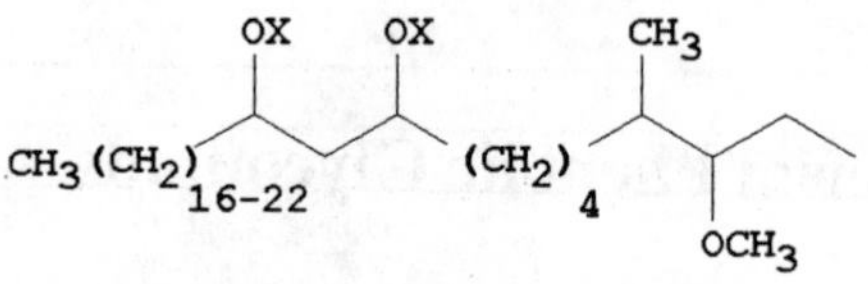

Phthiocerol B

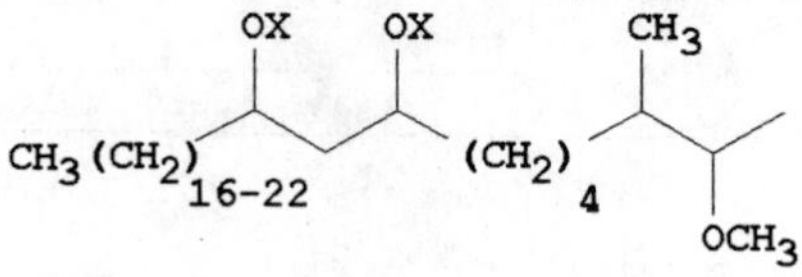

Phthiodiolone A

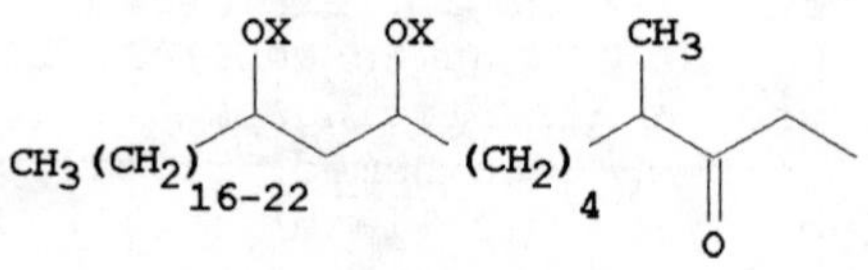

Phthiotriol A

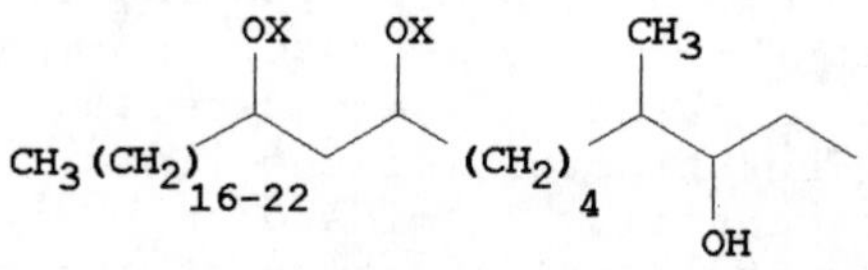

Phenolic glycolipid

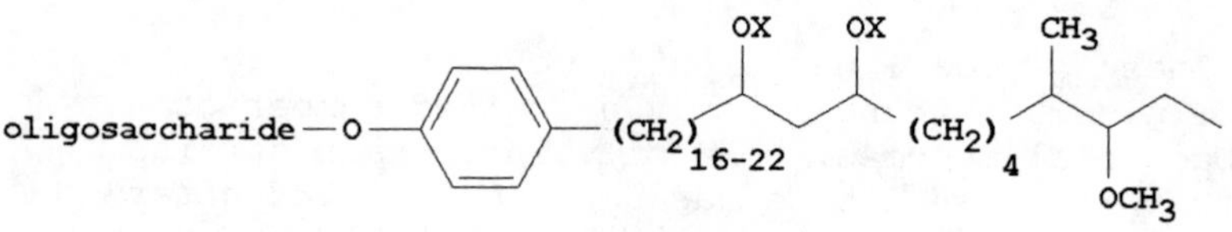

oligosaccharide — O

X = mycocerosate

$$CH_3(CH_2)_{16-20}\left[CH_2-\underset{\underset{CH_3}{|}}{CH}-CO\right]_{2-5}$$

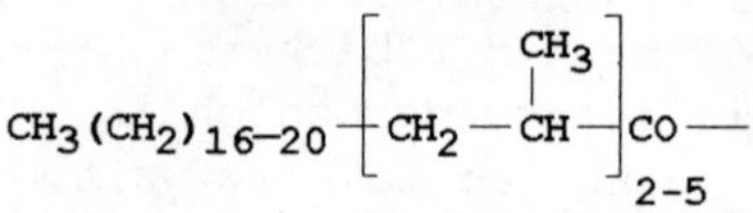

Figure 1 Structures of mycobacterial waxes and phenolic
          glycolipids

Table 1   Glycosyl units of the major phenolic
glycolipids

Taxon	Glycosyl units
M. kansasii *M. gastri*	2,6-dideoxy-4-*O*-methyl-arabinohexose, 2-*O*-methyl-fucose, 2-*O*-methyl-rhamnose, 2,4-*O*-methyl-rhamnose
M. bovis	2-*O*-methyl-rhamnose
M. leprae	3,6-*O*-dimethyl-glucose, 2,3-*O*-dimethyl-rhamnose, 3-*O*-methyl-rhamnose
M. tuberculosis Canetti strain	2,3,4-tri-*O*-trimethyl-fucose, rhamnose, 2-*O*-methyl-rhamnose
M. marinum *M. ulcerans*	3-*O*-methyl-rhamnose
M. haemophilum	2,3-*O*-dimethyl-rhamnose, 3-*O*-methyl-rhamnose, 2,3-*O*-dimethyl-rhamnose

between one and four sugars (Table 1). The β-diol unit,
in both phenolic glycolipids and phthiocerol-based
waxes, is esterified to two multi-methyl branched fatty
acids known as mycocerosic acids. Phthiocerol
dimycocerosates, however, have not shown any
antigenicity. Phenolic glycolipids and other
mycobacerial glycolipids have recently been reviewed by
Puzo[10], Brennan[11,12] and Minnikin.[13]

2 PHENOLIC GLYCOLIPIDS OF *MYCOBACTERIUM KANSASII* AND *MYCOBACTERIUM GASTRI*

The structure of the aglycone was first established by
Gastambide-Odier *et al.* by oxidative degradation of the
deacylated aglycone and analysis of the products by gas
chromatography and mass spectrometry.[8,9] For a long
time there was uncertainty over the sugar composition
of this glycolipid until the structure was assigned to
be 2,4-*O*-dimethyl-α-L-rhamnopyranosyl(1→ 4)2-*O*-methyl-fucopyranosyl(1→ 4)2-*O*-methyl-α-L-rhamnopyranosyl(1→)
phenolphthiocerol dimycocerosate.[14]

In recent years the structure of the phenolic glycolipid of *M. kansasii* (renamed Phe-Gl-K-I) has been reinvestigated and a minor component, Phe-Gl-K-II, was isolated by preparative high performance liquid chromatography (HPLC).[15] Fast atom bombardment (FAB) mass spectrometry experiments, with mono-butyl triethylene glycol as a matrix, showed molecular ions of a mass 186 Daltons higher than expected[16] and the ^{1}H-NMR spectrum indicated a tetrasaccharide structure with an acetate substituent on the carbohydrate moiety.[15]

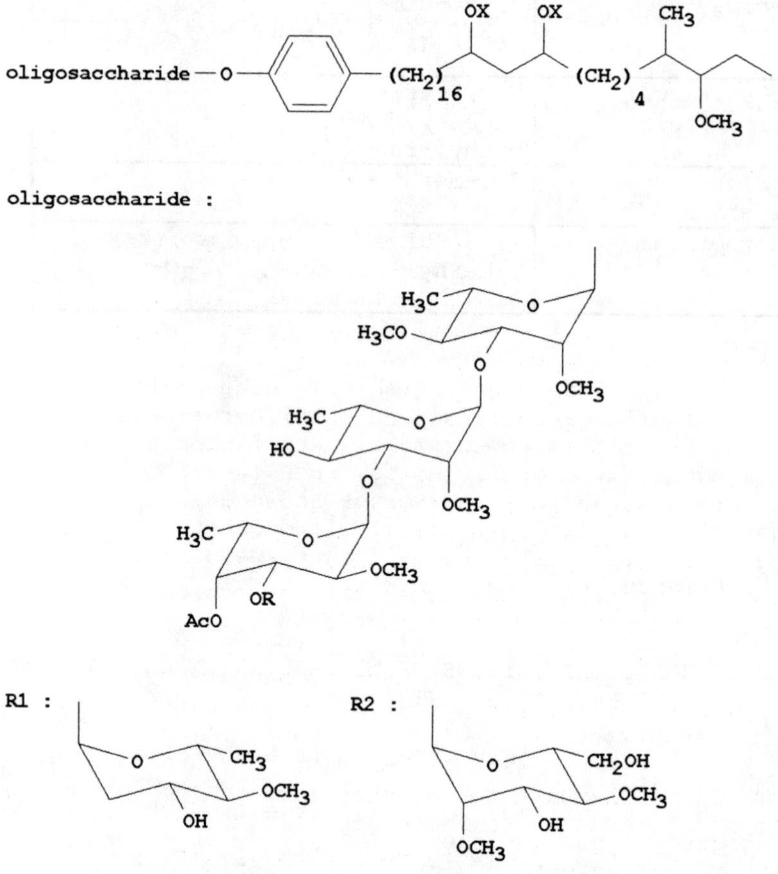

oligosaccharide :

R1 :   R2 :

Figure 2   The structures Phe-Gl-K-I and Phe-Gl-K-II, the major and minor phenolic glycolipids of *M. kansasii*. The substituent R corresponds to R1 and R2 for the major and minor lipids, respectively.

The additional sugar was determined to be the
acid-labile methyl 2,6-dideoxy-4-*O*-methyl-α-D-
arabinohexo-pyranoside. The total structure (Figure 2)
was established to be 2,6-dideoxy-4-*O*-methyl-α-D-
arabinohexopyranosyl(1→ 3)-2-*O*-methyl-4-*O*-acetyl-α-L-
fucopyranosyl(1→ 3)2-*O*-methyl-α-L-rhamnopyranosyl(1→
3)2,4-di-*O*-methyl-α-L-rhamnopyranosyl(1→)
phenolphthiocerol dimycocerosate.[17] The presence
of C_{29}, C_{30}, and C_{32} mycocerosic acids and a small
amount of glycolipids based on phenolphthiodiolone were
determined from the heterogeneity of the FAB mass
spectrum of Phe-Gl-K-I. The structure (Figure 2) of the
minor component Phe-Gl-K-II was determined in a similar
fashion, the only difference being the 2,4-di-*O*-methyl-
α-D-mannopyranoside sugar at the non-reducing end.[18] In
course of serological studies of the major phenolic
glycolipid, Phe-Gl-K-I, it was found that *Mycobacterium
gastri* produces an identical glycolipid.[59]

3 PHENOLIC GLYCOLIPIDS OF *MYCOBACTERIUM BOVIS*

Mycoside B was first discovered by Smith and Randall.[20]
Two groups investigated the structure of the lipid core
with the help of ^{1}H-NMR and mass spectroscopy.[8,9,21] It
was established to belong to the phenolphthiocerol
family. Mainly C_{29} mycocerosic acid was detected and
the saccharide was found to be 2-*O*-methyl-β-D-
rhamnose.[21] Recently the structure of this lipid has
been reinvestigated by Lanéelle *et al.*[22], Brennan *et
al.*[23], Puzo *et al.*[24] and Minnikin *et al.*[25] Desorption-
chemical ionization mass spectrometry (DCI-MS) with
ammonia as reagent gas confirmed the presence of C_{26},
C_{27}, C_{29} or C_{30} mycocerosic acids.[24] No evidence was
found for the presence of palmitic acid that had been
suggested in earlier reports. Gas chromatography of the
trimethylsilyl-2-L-butyl-glycosides of the 2-*O*-methyl-
rhamnose established its 'L' rather than 'D' absolute
configuration.[22,23] Carbon-13 NMR data indicated α-
linkage of the rhamnose to the phenol ring. The
structure of the major phenolic glycolipid (Phe-Gl-B)
of *M. bovis* is then 2-*O*-methyl-α-L-rhamnopyranosyl(1→)
phenolphthiocerol dimycocerosate (Figure 3). These
findings have been confirmed by two-dimensional ^{1}H-NMR
nuclear Overhauser effect spectrometry (NOESY).[26]

Three minor phenolic glycolipids (B-1, B-2,
B-3) have been isolated by HPLC and their structures
have been determined by DCI-MS, two dimensional ^{1}H-NMR

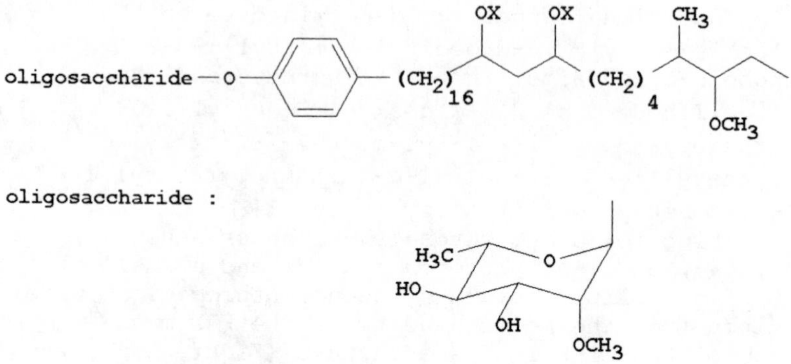

oligosaccharide :

Figure 3 Structure of Phe-Gl-B, the major phenolic
         glycolipid of *M. bovis*

homonuclear correlated spectroscopy (COSY) and NOESY
experiments. They vary either in the glycosyl moiety or
in the oxygenated functionality on the phenolphthiocerol unit as depicted in Figure 4.[26] Phe-Gl-B-1
and Phe-Gl-B-2 both have a phenolphthiodiolone
backbone. Phe-Gl-B-1 has the same glycosyl residue as
the major phenolic glycolipid Phe-Gl-B, whereas Phe-Gl-
B-2 carries a α-L-rhamnopyranosyl residue (Figure 4).

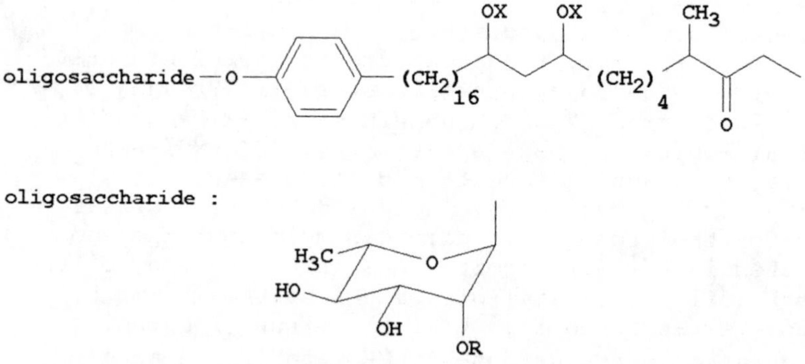

oligosaccharide :

R = CH$_3$ for Phe-Gl-B-1

R = H   for Phe-Gl-B-2

Figure 4 Structures of Phe-Gl-B-1 and Phe-Gl-B-2,
         minor phenolic glycolipids of *M. bovis*

In Phe-Gl-B-3 a phenolphthiocerol unit is combined with an α-L-rhamno-pyranosyl(1→3)2-*O*-methyl-α-L-rhamnopyranosyl glycosyl unit (Figure 5). There has also been a report of an additional minor phenolic glycolipid in *M. bovis* that from ^{1}H-NMR data seems to be a deacylated Phe-Gl-B.[23]

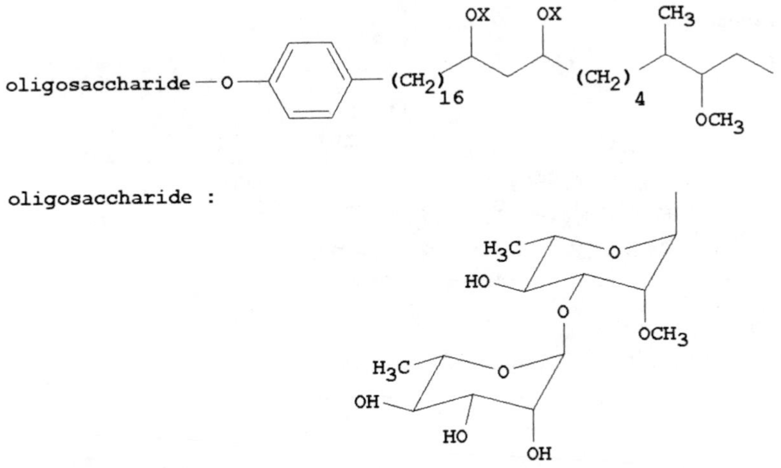
OX
OX
CH_3
oligosaccharide—O
$(CH_2)_{16}$
$(CH_2)_4$
OCH_3
oligosaccharide :
H_3C
O
HO
O
OCH_3
H_3C
O
OH
HO
OH

Figure 5 Structure of Phe-Gl-B-3, a minor phenolic glycolipid of *M. bovis*

A phenolic glycolipid, co-chromatographing with the major lipid from *M. bovis*, was found in representatives of *Mycobacterium africanum* and *Mycobacterium microti*.[27] These species are closely related to *M. bovis* and *M. tuberculosis*, as members of the 'tuberculosis complex'.

4 PHENOLIC GLYCOLIPIDS OF *MYCOBACTERIUM LEPRAE*

Following the availability of larger amounts of *M. leprae* from infected armadillo tissue, two groups independently found evidence for a phenolic glycolipid called PGL-I.[28,29] It was found to be identical with a previously detected lipid antigen of *M. leprae*.[30] A review by Brennan summarizes early work on *M.leprae* surface antigens.[31] Proton and ^{13}C-NMR studies confirmed the carbohydrate structure and the anomeric configuration and the structure is established to be 3,6-*O*-dimethyl-β-D-glucopyranosyl(1→4)-*O*-2,3-dimethyl-α-L-rhamnopyranosyl(1→2)-3-*O*-methyl-α-L-rhamnopyranosyl(1→) phenolphthiocerol dimycocerosate

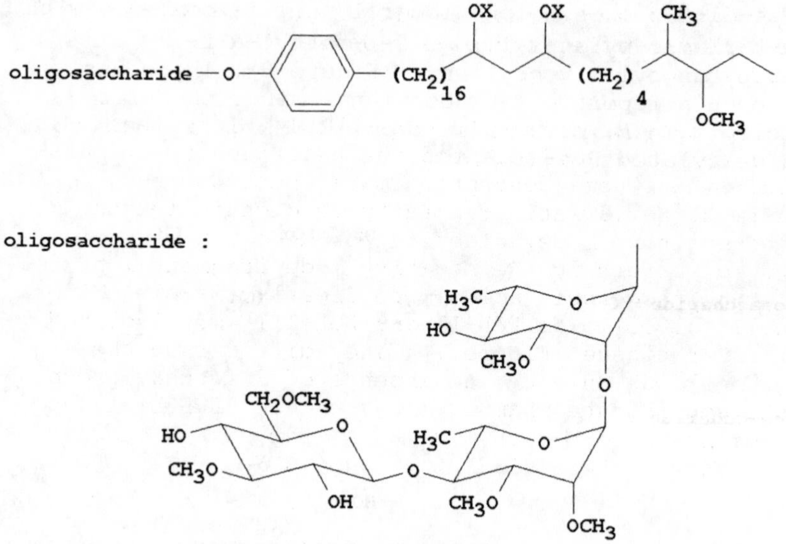

Figure 6 Structure of PGL-I, the major phenolic
         glycolipid of *M. leprae*

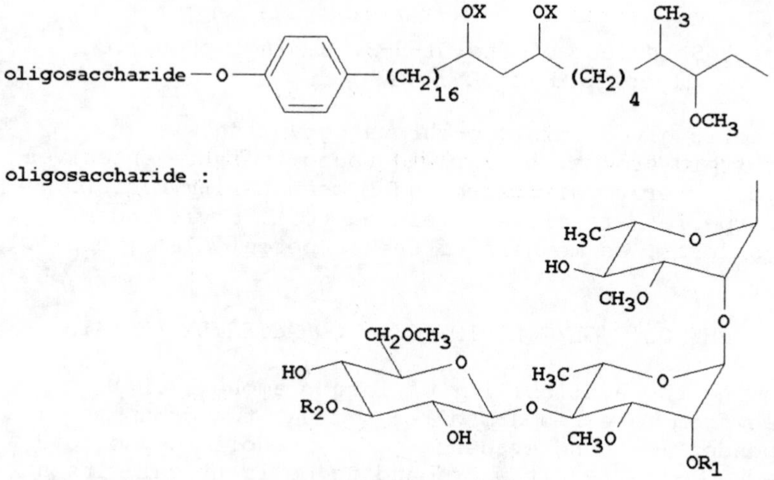

PGL-II :   R = H R_2 = CH_3

PGL-III : R_1=CH_3 R_2 = H

Figure 7 Structures of PGL-II and PGL-III, the minor
         phenolic glycolipids of *M. leprae*

(Figure 6).[32] Mycocerosates with 30,32 and 34 carbons were detected after hydrolysis and GC/MS of the methyl esters.

FAB mass spectra with mono-butyl triethylene glycol as a matrix indicated the heterogenity of this lipid[16] and plasma desorption mass spectrometry confirmed the variation in fatty acid chain length.[33] In the positive mode, clusters for M+Na ion were detected whereas in the negative mode fragment ions from the mycocerosic acids were clearly visible. Minor phenolic glycolipids PGL-II and PGL-III were isolated from liver tissue (Figure 7). They differ from the major analogue only by the absence of 2-methoxy group at the penultimate rhamnose in the case of PGL-II and the absence of 3-*O*-methyl group at the distal glucose in the case of PGL-III.[34]

5 PHENOLIC GLYCOLIPIDS OF *MYCOBACTERIUM TUBERCULOSIS*

A phenolic glycolipid was recently discovered in the unusual smooth Canetti strain of *M. tuberculosis*. These are the only strains of *M. tuberculosis* that produce

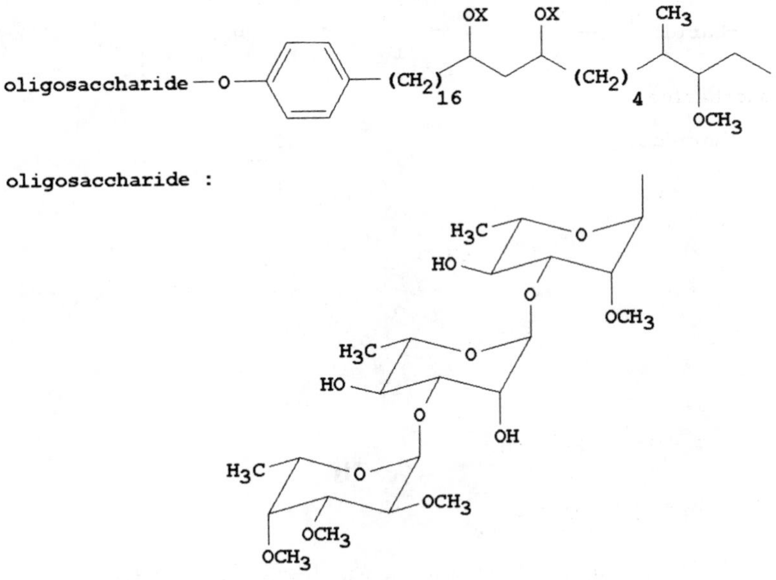

Figure 8 Structure of PGL-Tb-I, the major phenolic glycolipid of *M. tuberculosis* Canetti

large amounts of a phenolic glycolipid. The structure
(Figure 8) of the major component, PGL-Tb-I, was
determined to be 2,3,4-tri-*O*-methyl-α-L-
fucopyranosyl(1→3)α-L-rhamno-pyranosyl(1→3)2-*O*-
methyl-α-L-rhamnopyranosyl(1→)phenolphthiocerol
dimycocerosate.[35] A further structural study of this
lipid by various 2D-NMR techniques confirmed the
structure and with a J-resolved spectrum, the presence
of two minor components was detected. These vary in the
non-saccharide end of the molecule, carrying a keto-
rather than a methoxy group, or lacking a methylene
group at the lipophilic end of the phenolphthiocerol.[36]

    Small amounts of the major phenolic glycolipid of
M. bovis BCG Phe-Gl-B (Figure 3) were also detected in
M. tuberculosis Canetti strain. There is a strong
sructural similarity between PGL-Tb-I and one of the
minor phenolic glycolipids of *M. bovis*. Phe-Gl-B-3
(Figure 5) only differs in the absence of the distal
tri-*O*-methyl-fucopyranose moiety.

    A further minor component was isolated and its
structure was established from NMR data of the native

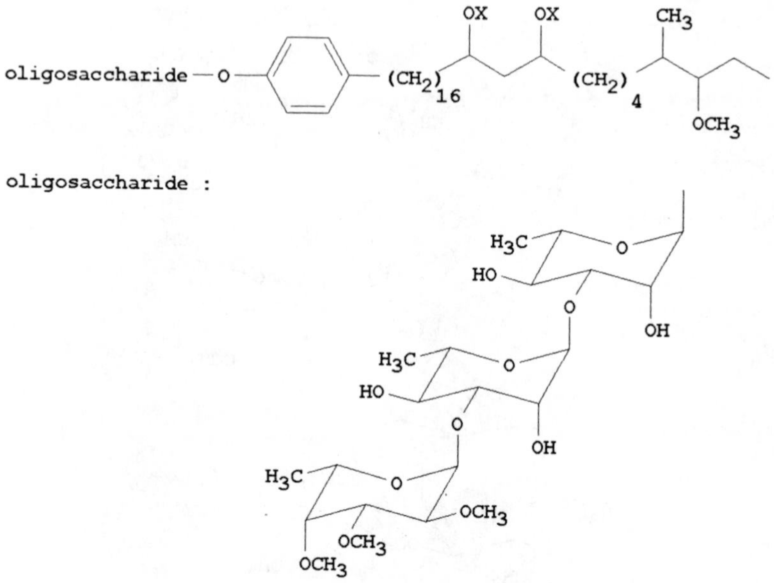

oligosaccharide :

Figure 9 Structure of the minor phenolic glycolipid of
         M. tuberculosis Canetti varying in the
         oligosaccharide moiety

lipid and its peracetylated derivative. The structure (Figure 9) was found to be 2,3,4-tri-*O*-methyl-α-L-fucopyranosyl(1→3) α-L-rhamnopyranosyl(1→3) α-L-rhamnopyranosyl(1→)phenolphthiocerol dimycocerosate.[37] As it shares the two distal sugars with the antigenic major phenolic glycolipid of *M. tuberculosis* Canetti strain it is likely to be antigenic. It has been suggested that this minor component is a precursor in the biosynthesis of PGL-Tb-1.

6 PHENOLIC GLYCOLIPIDS OF *MYCOBACTERIUM MARINUM* AND *MYCOBACTERIUM ULCERANS*

Mycoside G from *M. marinum* strain Balnei was first discovered by Smith.[38] The structure (Figure 10) was established to be 3-*O*-methyl-α-L-rhamnospyranosyl(1→) phenolphthiocerol dimycocerosate.[39,40] Its sugar unit differs from the phenolic glycolipid of *M. bovis* only in the position of methoxy group on the glycosyl residue. The main mycocerosic acids, found esterified to the diol unit, had 24, 27 and 30 carbons.

Minor components with methyl branches on the phenolphthiocerol unit and phenolphthiodiolone units have been reported as well as a phenol phthiotriol lipid core.[27,41] There are still uncertainties about the existence of another minor phenolic glycolipid called mycoside G' which has been reported to have a mycolic acid esterified to the 2-hydroxyl group of the 3-*O*-methyl rhamnopyranose.[42]

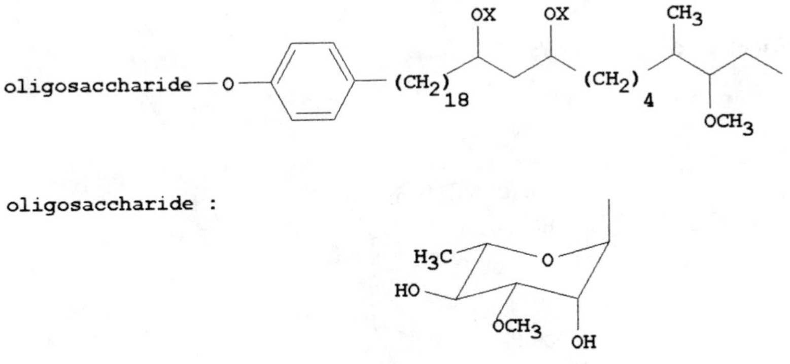

Figure 10 Structure of the major phenolic glycolipid of *M. marinum*

A diacyl phenolphthiocerol was first detected in *M. ulcerans* in 1984.[43,44] A stereochemical study established its relation to phenolphthiocerols of *M. marinum.*[45] A phenolic glycolipid was detected in *M. ulcerans* recently.[45] From NMR studies and degradation experiments the structure was shown to be identical with that of the phenolic glycolipid (mycoside G) of *M. marinum.*[46] However this lipid was only detected in 2 of 20 strains examined. The main mycocerosic acid was found to be C_{27}.

7 PHENOLIC GLYCOLIPID OF *MYCOBACTERIUM HAEMOPHILUM*

A phenolic glycolipid was first detected in *M. haemophilum* in 1990.[47] Its antigenicity towards a serum raised against whole cells was demonstrated by an enzyme-linked immunosorbent assay (ELISA) test. The structure (Figure 11) was elucidated with the help of ^{1}H-NMR and COSY spectra and the usual degradation experiments. The structure was determined to be 2,3-di-*O*-methyl-α-L-rhamnopyranosyl(1→2) 3-*O*-methyl-α-L-rhamnopyranosyl(1→4) 2,3-di-*O*-methyl-α-L-rhamnopyranosyl(1→) phenolphthiocerol dimycocerosate.[48] The mycocerosates found were C_{27}, C_{30}, C_{32}, C_{34}, and C_{37}.

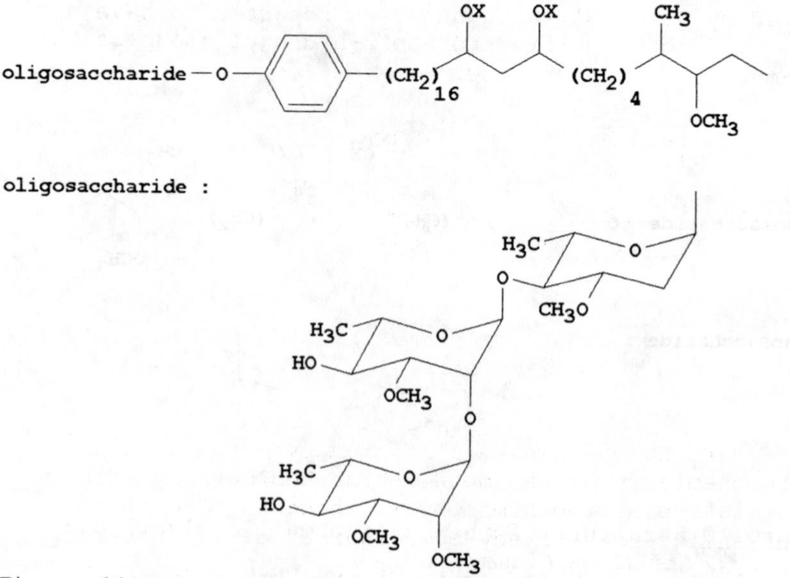

Figure 11 Structure of the major phenolic glycolipid of *M. haemophilum*

It is the only phenolic glycolipid apart from the PGL-I of *M. leprae* that contains the tetramethyl C_{34} mycocerosic acid. A further similarity to the *M. leprae* phenolic glycolipid is the number of carbons in the lipid moiety and the occurrence of the same two internal saccharides linked in inverse order.

8 STEREOCHEMICAL STUDIES ON PHENOLIC GLYCOLIPIDS

The main biological activity of the phenolic glycolipids lies in the oligosaccharide core of these molecules. The lipid core is thought to mainly provide the anchorage of these surface lipids in the mycobacterial cell wall.[12,13,49,50,51] There is an interest however in the stereochemistry of the four chiral carbon centres in the phenolphthiocerol backbone and the asymmetric carbon atoms that carry the methyl branches in the mycocerosic acids.

The first investigation in this direction was published by Polgar and co-workers.[52] By comparison of ^{1}H-NMR data and IR spectra of the native phthiocerol of *M. tuberculosis* and synthetic analogues (*threo*- and *erythro*-decane-4,6-diols) and the corresponding ethylidene acetals, it was shown that the β-diol unit of the phthiocerol of *M. tuberculosis* has a *threo* relative stereochemistry. The absolute stereochemistry of the diol unit was later found to be *R, R* (Figure 12) from comparison of optical rotations of a series of cyclic 1,3-dioxanes.[53]

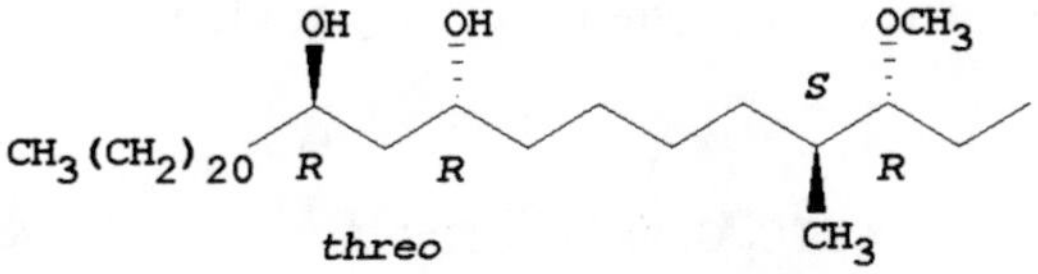

Figure 12 Absolute stereochemistry of phthiocerol A
         of *M. tuberculosis*

There has been some confusion over the absolute stereochemistry of the methyl-branched methoxy unit. The relative stereochemistry is thought to be *threo* after an intensive study by Polgar et al..[54] After reduction of the β-diol unit of the phthiocerol, ^{1}H-NMR data were compared to that of the model compound, *erythro*-2-methoxy-3-methylhentriacontane. The absolute

stereochemistry (Figure 12) was elucidated soon
after.[55] By reductive removal of the methoxy- and β-
diol group and comparison of the optical rotations the
methyl branch was found to have the *S*- and the methoxy
branch the *R*-configuration. Similar assignments have
not been made for any of the phenolphthiocerols.

In recent years several attempts were made to
investigate the relative stereochemistry of the β-diol
unit of phthiocerols and phenolphthiocerols of various
mycobacterial species. By IR and ^{1}H-NMR spectroscopy it
was shown that *M. marinum*[56] and *M. ulcerans*[45] differ
from all other mycobacteria which produce phthiocerols
and phenolphthiocerols by having an *erythro*
configuration of the β-diol unit rather than *threo*.[57]
From optical rotation measurements it has been
suggested that *M. marinum* and *M. ulcerans* also have a
different stereochemistry in the methyl-branched
methoxy unit.[58] Moreover, contrary to all other
mycobacteria, *M. marinum* and *M. ulcerans* have
dextrorotatory mycocerosic acids esterified to
phthiocerols and phenolphthiocerols.[44] .

The stereochemical difference between the phenolic
glycolipids from *M. marinum* and *M. ulcerans* and those
from all other mycobacteria can be readily determined
by simple inspection of the ^{1}H-NMR spectra. The signal
for the >CH-OX units (Figures 1 to 11) is at δ 4.90 ppm
for the lipids from *M. marinum* and *M. ulcerans* and at δ
4.84 ppm for phenolic glycolipids from other
mycobacteria.[25,41,45,48,56] It appears, therefore, that
M. marinum and *M. ulcerans* have chemotaxonomically
significant differences in their phenolic glycolipid
structures.

9 ANTIGENICITY OF PHENOLIC GLYCOLIPIDS AND
  SYNTHETIC ANALOGUES

It has been shown, that the phenolic glycolipids of *M.
kansasii,* Phe-Gl-K-I and Phe-Gl-K-II, react against
polyclonal antibodies from rabbits, inoculated with
killed *M. kansasii* cells.[17] Both enantiomers of the
distal methyl 2,6-dideoxy-4-*O*-methyl-α-arabino-
hexopyranose have been synthesized by two groups[59,60]
and only the D-enantiomer and the mono-acetyl
triglycosyl Phe-Gl-K-I inhibited the linkage of the
Phe-Gl-K-I to Phe-Gl-K-I antibodies in an ELISA
inhibition experiment. The other di- and tri-saccharide
Phe-Gl-K-I and the deacylated lipid exhibited no

antigenicity. Thus the epitope of this antigen was determined to be the distal mono-acetylated disaccharide.[59] In the course of these ELISA studies it was found that *M. gastri* produces an identical antigenic phenolic glycolipid.[19]

In 1982 Payne *et al*.[29] first demonstrated the antigenicity of PGL-I by reacting it against undiluted sera of lepromatous leprosy patients. Three groups suggested different techniques to overcome problems arising from the insolubility of this lipid in aqueous buffers, hampering ELISA experiments. Brennan *et al*.[61] used a sonicated emulsion of PGL-I, Young and Buchanan[62] deacylated the lipid to increase its polarity and Brett *et al*.[63] suggested the use of detergent. PGL-I proved to be a serological marker for lepromatous leprosy with over 90% of clinically diagnosed patients having high antiglycolipid IgM immunoglobulin titres.[61] However for the less easily diagnosed tuberculoid leprosy these assays were insensitive.

Fujiwara defined the PGL-I epitope by synthesizing the mono-, di-, and trisaccharide.[64,65] The distal monosaccharide unit was shown to be immunodominant by ELISA inhibition technique but the efficiency of the inhibitor is increased by the presence of the distal disaccharide. Alternative synthethic strategies towards the di- and trisaccharide were published by Gigg *et al*.[66,67] and Verez-Bencomo *et al*. [68,69] Young *et al*.[70] supported the findings on the antigenicity by raising 9 monoclonal antibodies out of which 5 recognized selectively the distal glucose.

Neoantigens have been synthezised by attaching the di- and trisaccharide to bovine serum albumin (BSA) as a carrier protein. Different linker arms have been used for this purpose :
- (8-methoxycarbonyl)octyl group,[71-73]
- p-(2-methoxycarbonylethyl)phenyl group,[74]
- allyl group.[67,75]

The di- and trisaccharide neoantigens used with the first linker arm showed similar sensitivity as the trisaccharide neoantigen containing the phenol linker in ELISA tests against sera of lepromatous leprosy patients.[73] None of the synthetic neoanitigen exhibited higher sensitivity or specificity than the native PGL-I. For this reason and their increased solubility in aqueous solutions they have been established as a good and cheap alternative to the native lipid for the serodiagnosis of leprosy.

An alternative strategy in the synthesis of neoantigens was provided by preparing copolymers of the trisaccharide attached to an allyl group and acrylamide. They have the same sensitivity as the protein linked neoantigens but a lower specificity.[70] Brennan *et al.*[76] have developed a method for the quantification of PGL-I which is of particular value as the amount of PGL-I during chemotherapy decreases faster than the level of the antibodies.

Preliminary studies have shown that antibodies reacting against PGL-Tb-I are present in sera of most tuberculosis patients.[77] The use of PGL-Tb-I in the serodiagnosis of tuberculosis remains an attractive proposition, despite the fact that this lipid apparently only occurs in substantial amounts in the smooth Canetti variants of *M. tuberculosis*. Limited evidence has been advanced for the presence of PGL-Tb-I in recent isolates of *M. tuberculosis*[78] and this has been reinforced by serological evidence.[79] In addition to PGL-Tb-I, smooth Canetti strains also produce a characteristic antigenic lipooligosaccharide[80,81] and in rough and smooth strains a family of diacyl trehalose antigens are present.[82,83,84,85,86,87] Comparison of the value of these three types of lipids in the serodiagnosis of tuberculosis clearly demonstrated that no single lipid detects all cases but a combination is most effective.[83,86]

Fujiwara successfully synthesized the saccharide moiety of PGL-Tb-I and linked it to BSA via a p-(2-methoxycarbonylethyl)phenyl linker.[88] This neoantigen showed good activity and specificity against human tuberculosis sera. Even better results were obtained by linking the trisaccharide to hemocyanin from 'keyhole limpets'.[88]

Mycoside B and the minor phenolic glycolipids of *M. bovis* have reacted positively in ELISA experiments against anti-*M. bovis* BCG rabbit polyclonal antibodies.[26] Furthermore Phe-Gl-B and Phe-Gl-B-3 are immunogenic and rabbit polyclonal antibodies against these two lipids were raised. A neoantigen was synthesized by Brennan containing 2-*O*-methyl-α-L-rhamnose and (8-methoxycarbonyl)octyl group as a spacer.[23] This neoantigen seems to be specific to *M. bovis* but as mycoside B does not occur in several *M. bovis* strains it is only of limited use as a serological marker.

The phenolic glycolipid of *M. haemophilum* has been shown to react in ELISA tests with antisera raised against whole cells of *M. haemophilum*.[47] No attempts have yet been made to synthesize its oligosaccharide unit and to study the antigenicity of the saccharides or possible neoantigens.

The phenolic glycolipid of *M. marinum* (mycoside G) was shown to be a specific antigen, reacting strongly with homologous antisera prepared from whole organisms.[83] Attempts to raise antibodies in rabbits by direct injection of mycoside G were unsuccessful.[89]

10 CONCLUSIONS

Phenolic glycolipids are unusual surface-active lipids found only in a very limited range of pathogenic mycobacteria (Table 1). Compared with other surface-active lipids, phenolic glycolipids are extremely hydrophobic, having a high portion of long-chain components and relatively hydrophobic sugars. The biological activity of these lipids depends on their lipid chains being bound with other long chains, exposing the antigenic sugar epitope on the cell surface. Following an original proposal,[49] a chemical model of the mycobacterial cell wall (Figure 13) has been developed.[11,13,50,51] It was considered,[49] that the mycocerosate chains of the phenolic glycolipids (Figures 1 to 11) and waxes based on the phthiocerol family (Figure 1) are designed to interact specifically with high molecular weight mycolic acids covalently bound to the cell wall. The phthiocerol-based waxes would have a role as 'inert fillers' with the 'active' phenolic glycolipids interspersed over the cell surface. Phenolic glycolipids are present in both rough and smooth colony variants of *M. kansasii*, though the rough variants lack a family of characteristic antigenic lipooligosaccharides.[90]

The specific antigenicity of phenolic glycolipids has been summarized above but it is intriguing that apparently different mycobacterial species produce the same lipid antigen and certain species have a heterogenous distribution of such lipids. For example, *M. marinum* and *M. ulcerans* can synthesize the same phenolic glycolipid though they cause very different diseases. Similarily the pathogen *M. kansasii*, and the very attenuated *M. gastri* produce phenolic glycolipids which are practically identical. These relationships raise important phylogenetic questions.

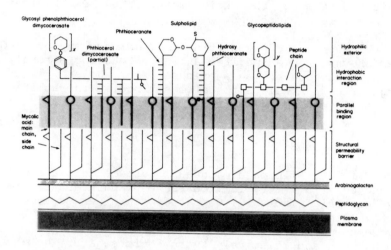

Figure 13 Chemical model of the lipid interaction in
           the cell envelope of mycobacteria. The model
           shows how various complex lipids such as the
           dimycocerosates of glycosyl phenol-
           phthiocerols (Figures 1 to 10) and the
           phthiocerol family (Figure 1) may interact
           with the mycolic acid chains. Reproduced
           with permission from Reference 13.

     The distribution of phenolic glycolipids in the
'tuberculosis complex' poses other questions. Three of
the four species, *M. africanum*, *M. bovis*, and *M.
microti* produce mainly Phe-Gl-B (Figure 3) but only a
few unusual smooth Canetti strains of *M. tuberculosis*
synthesize this lipid and an additional more polar
phenolic glycolipid, PGL-Tb-I (Figure 8). Members of
the 'tuberculosis complex' are genetically almost
identical even though their virulence is different. It
is clear that further studies are necessary to clarify
the precise relationships between members of the
'tuberculosis complex'.

     The most surprising phenomenon concerns the
stereochemical differences between the phenolic
glycolipids of *M. marinum* and *M. ulcerans* and those

from other mycobacteria (Sections 6 and 8).
Superficially, all the lipids are very similar but the
β-diol unit in *M. marinum* and *M. ulcerans* have *erythro*
configuration in contrast to the *threo* arrangement for
those from the other phenolic glycolipids. In parallel,
the mycocerosates from the two groups have opposite
absolute configuration. Presumably, the resulting
overall conformation of these molecules must be
comparable so they can interact in the same way with
the mycolic acid chains in the mycobacterial cell
envelope (Figure 13). The general biochemical pathways
in all these mycobacteria appear to be rather similar,
so at some stage in the past the biosynthesis of
phenolic glycolipids in *M. marinum* and *M. ulcerans* must
have diverged from that leading to the lipids in the
other mycobacterial species.

The heterogeneity of the distribution of phenolic
glycolipids (Table 1) and their unique structures gives
these lipids great potential in identifying the parent
species. A very simple procedure for the extraction of
phenolic glycolipids, involving brief shaking with a
biphasic mixture of petroleum ether and aqueous
methanol,[27,91] followed by simple thin-layer
chromatographic analysis of the extracts, allows the
identification of all the phenolic glycolipids,
excepting the relatively simple lipids from *M. bovis*
(Figure 3) and *M. marinum* (Figure 10). The sugar units
in these latter lipids differ only in the position of a
methyl group but on spraying the TLC plate with α-
naphthol/sulphuric acid, followed by heating, very
distinct colours are produced.[27,90] By such simple
means an unknown mycobacterial isolate can be
identified very rapidly with good precision. Further
information of taxonomic value can be obtained by
degrading phenolic glycolipids and analysing the
constituent sugars[27] or long-chain components.[41]

Direct analysis of specific lipids in infected
tissue can produce direct diagnosis without prior
cultivation of the causative organism. The mycocerosic
acids, from the phenolic glycolipids, give
characteristic profiles and they are easily released by
alkaline hydrolysis and converted to pentafluorobenzyl
esters. These esters can be detected down to picomole
or femtomole levels by the use of either electron
capture gas chromatography(EC-GC) or negative ion
chemical ionisation mass spectrometry(NI-CI GC-MS).[92]
In a preliminary study it was shown that mycocerosate
profiles, characteristic of *M. leprae*, could be easily
detected in leprosy skin biopsies by EC-GC.[93] It was

later shown[94], that NI-CI GC-MS of mycocerosate
pentafluorobenzyl esters[91] was a highly sensitive way
of detecting the lipids of *M. leprae* in infected
tissue.

ACKNOWLEDGEMENTS

Studentships from the British Leprosy Relief
Association to S. Hartmann and G.S. Besra and G. Dobson
have supported the work on the phenolic glycolipids of
the leprosy bacillus and related mycobacteria. A grant
from the World Health Organisation Programme for
Vaccine Developement, to D.E. Minnikin, has supported
studies on the phenolic glycolipid of *M. tuberculosis*
Canetti. A Medical Research Council project grant
(G8216538), to D.E. Minnkin and M. Goodfellow, enabled
analyses of phenolic glycolipid profiles to be
recorded. A grant from the Wellcome Trust
(030923/Z/89/Z/JGH/CDJS/EO), to D.E. Minnikin, has
supported work on the detection of characteristic
lipids in infected tissue.

REFERENCES

1. D.W. Smith, W.K. Harrell, H.M. Randall, Am.
    Rev.Tuberculosis, 1954, 66, 505.
2. D.W. Smith, H.M. Randall, A.P. MacLennan, E.
    Lederer, Nature, 1960, 186, 887.
3. A.P. MacLennan, H.M. Randall, Biochem. J., 1961,
    80, 309.
4. A.P. MacLennan, D.W. Smith, H.M. Randall, Biochem.
    J., 1960, 74, 3.
5. D.W. Smith, H.M. Randall, A.P. MacLennan, R.K.
    Putney, S.V. Rao, J. Bacteriol., 1960, 79, 217.
6. D.W. Smith, H.M. Randall, M. Gastambide-Odier,
    A.L. Koevoet, Ann. N. Y. Acad. Sci., 1957,69, 145.
7. H. Noll, J. Biol. Chem., 1957, 224, 149.
8. M. Gastambide-Odier, P. Sarda, Pneumologie, 1970
    142, 241.
9. M. Gastambide-Odier, P. Sarda, E. Lederer, Tet.
    Lett., 1965, 35, 3135.
10. G. Puzo, Crit. Rev. Microbiol., 1990, 17, 305.
11. P.J. Brennan, Rev. Infect. Dis, 1989, 2
    (suppl.2), S420
12. P.J. Brennan in 'Microbial Lipids', Ed. C
    Ratledge, S.G. Wilkinson, Academic Press,
    London, 1988, p. 203.

13. D.E. Minnikin in 'Chemotherapy of Tropical Diseases', Ed. M. Hooper, Wiley, Chichester, 1987, p. 19.

14. V. Mehra, P.J. Brennan, E. Rada, J. Convit, B.R. Bloom, Nature, 1984, 308, 194.

15. J.J. Fournié, M. Rivière, G. Puzo, J. Biol. Chem., 1987, 262, 3174.

16. M. Rivière, J.J. Fournié, A. Vercellone, G. Puzo, Biomed. Environ. Mass Spectrom., 1988, 16, 275.

17. J.J. Fournié, M. Rivière, F. Papa, G. Puzo, J. Biol. Chem., 1987, 262, 3180.

18. M. Rivière, J.J. Fournié, G. Puzo, J. Biol. Chem., 1987, 262, 14879.

19. F. Papa, M. Rivière, J.J. Fournié, G. Puzo, H. David, J. Clin. Microbiol., 1987, 25, 2270.

20. H.M. Randall, D.W. Smith, J. Opt. Soc. Am., 1953, 43, 1086.

21. E. Lederer, H. Demarteau-Ginsburg, Biochim. Biophys. Acta, 1963, 70, 442.

22. M. Daffé, M.A. Lanéelle, C. Lacave, M.A. Lanéelle, Biochim. Biophys. Acta, 1988, 958, 443.

23. D. Chatterjee, C.M. Bozic, C. Knisley, S.N. Cho, P.J. Brennan, Infect. Immun., 1989, 57, 322.

24. A. Vercellone, M. Rivière, G. Puzo, Adv. Mass Spectrom., 1989, 11B, 1350.

25. D.E. Minnikin, G. Dobson, R.C. Bolton, J.H. Parlett, A.I. Mallet in 'Analytical Microbiology Methods : Chromatography and Mass Spectrometry, Ed. A. Fox, S.L. Morgan, L. Larsson, G. Odham, Plenum Press, New York, 1990, p. 137.

26. A. Vercellone, G. Puzo, J. Biol. Chem., 1989, 264, 7447.

27. D.E. Minnikin, G. Dobson, J.H. Parlett, M. Goodfellow, M. Magnusson, Eur. J. Clin. Microbiol., 1987, 6, 703.

28. S. Hunter, P.J. Brennan, J. Bacteriol., 1981, 147, 728.

29. S.N. Payne, P. Draper, R.J.W. Rees, Int. J. Lepr., 1982, 50, 220.

30. P.J. Brennan, W.W. Barrow, Int. J. Lepr., 1981, 48, 382.

31. P.J. Brennan, Int. J. Lepr., 1983, 51, 387.

32. E. Tarelli, P. Draper, S.N. Payne, Carbohydr. Res., 1984, 131, 346.

33. I. Jardine, G. Scanlan, M. McNeil, P.J. Brennan, Anal. Chem., 1989, 61, 416.

34. S. Hunter, P.J. Brennan, J. Biol. Chem., 1987, 263, 7556.

35. M. Daffé, C. Lacave, M-A. Lanéelle, G. Lanéelle, Eur. J. Biochem., 1987, 167, 155.

36. M. Daffé, P. Servin, Eur. J. Biochem., 1989, 185,
     157.
37. M. Daffé, Biochim. Biophys. Acta, 1989, 1002, 257.
38. R.G. Navalkar, E. Wiegeshaus, E. Kondo. H.K. Kim,
     D.W. Smith, J. Bacteriol., 1965, 90, 262.
39. M. Gastambide-Odier, C. Villé, Carbohydr. Res.,
     1970, 12, 97.
40. P. Sarda, M. Gastambide-Odier, Chem. Phys. Lipids,
     1967, 434.
41. G. Dobson, D.E. Minnikin, G.S. Besra, A.I. Mallet,
     M. Magnuson, Biochim. Biophys. Acta, 1990, 1042,
     176.
42. M. Gastambide-Odier, Eur. J. Biochem., 1973, 33,
     81.
43. M. Daffé, M.A. Lanéelle, J. Roussel, C.
     Asselineau, Ann. Inst. Pasteur/Microbiol., 1984,
     135A, 191.
44. M. Daffé, M.A. Lanéelle, J. Gen. Microbiol., 1988,
     134, 2049.
45. G.S. Besra, D.E. Minnikin, A. Sharif, J.L.
     Stanford, FEMS Microbiol. Lett., 1990, 66, 11.
46. M. Daffé, A. Varnerot, V.V. Lévy-Frébault, J. Gen.
     Microbiol., 1992, 138, 131.
47. G.S. Besra, D.E. Minnikin, L. Rigouts, F.
     Portaels, M. Ridell, Let. Appl. Microbiol., 1990,
     11, 202.
48. G.S. Besra, M. McNeil, D.E. Minnikin, F. Portaels,
     M. Ridell, P.J. Brennan, Biochemistry, 1991, 30,
     7772.
49. D.E. Minnikin in 'The Biology of the
     Mycobacteria', Ed. C. Ratledge, S.G. Wilkinson,
     Academic Press, London, 1982, p. 95.
50. P.J. Brennan, Res. Microbiol., 1991, 142, 451.
51. D.E. Minnikin, Res. Microbiol., 1991, 142, 423.
52. K. Maskens, D.E. Minnikin, N. Polgar, J. Chem.
     Soc., 1966, 2113.
53. J.-F. Tocanne, Bull. Soc. Chim. France, 1970, 750
54. K. Maskens, N. Polgar, J. Chem. Soc. Perkin I,
     1973, 1117.
55. K. Maskens, N. Polgar, J. Chem. Soc. Perkin I,
     1973, 1909.
56. G.S. Besra, A.I. Mallet, D.E. Minnikin, M. Ridell,
     J. Chem. Soc. Chem. Commun., 1989, 1451.
57. G.S. Besra, D.E. Minnikin, M. Ridell, M.
     Magnusson, Health Cooperation Papers, 1992, 12,
     45.
58. M. Daffé, Res. Microbiol., 1991, 142, 405.
59. M. Gilleron, J.J. Fournié, J.R. Pougny, G, Puzo,
     J. Carbohydr. Chem., 1988, 7, 733.
60. M.K. Gurjar, P.K. Ghosh, J. Carbohydr. Chem.,
     1988, 7, 799.

61. S.N. Cho, D.L. Yanagihara, S.W. Hunter, R.H. Gelber, P.J. Brennan, Infect. Immun., 1983, 41, 1077.

62. D.B. Young, T.M. Buchanan, Science, 1983, 221, 1057.

63. S.J. Brett, P. Draper, S.N. Payne, R.J.W. Rees, Clin. Exp. Immunol., 1983, 52, 271.

64. T. Fujiwara, S.W. Hunter, S.N. Cho, G.O. Aspinall, P.J. Brennan, Infect. Immun., 1984, 43, 245.

65. T. Fujiwara, S.W. Hunter, P.J. Brennan, Carbohydr. Res., 1986, 148, 287.

66. J. Gigg, R. Gigg, S. Payne, R. Conant, Chem. Phys. Lipids, 1985, 38, 299.

67. R. Gigg, S. Payne, R. Conant, J. Carbohydr. Chem., 1983, 2, 207.

68. J. Marino-Albernas, V. Verez-Bencomo, L. Gonzales, C.S. Perez, Carbohydr. Res., 1987, 165, 197.

69. J. Marino-Albernas, V. Verez-Bencomo, L. Gonzales-Rodriguez, C.S. Perez-Martinez, E. Gonzalez-Abreu, A. Gonzalez-Segredo, Carbohydr. Res., 1988, 183, 175.

70. D.B. Young, S.R. Khanolkar, L.L. Barg, T.M. Buchanan, Infect. Immun., 1984, 43, 183.

71. D. Chatterjee, J.T. Douglas, S.N. Cho, T. H. Rhea, R.H. Gelber, G.O. Aspinall, P.J. Brennan, Glycoconjugate J., 1985, 2, 187.

72. D. Chatterjee, S.N. Cho, P.J. Brennan, Carbohydr. Res., 1986, 156, 39.

73. D. Chatterjee, S.N. Cho, J.T. Douglas, T. Fujiwara, P.J. Brennan, Carbohydr. Res., 1988, 183, 241.

74. T. Fujiwara, S. Izumi, Agric. Biol. Chem., 1987, 51, 2539.

75. S.J. Brett, S.N. Payne, J. Gigg, P. Burgess R. Gigg, Clin. Exp. Immunol., 1986, 64, 476.

76. S.N. Cho, S.N. Hunter, R.H. Gelber, T.H. Rea, P.J. Brennan, J. Infect. Dis., 1986, 153, 560.

77. J. Torgal-Garcia, H.L. David, F. Papa, Ann. Inst. Pasteur/Microbiol., 1988, 139, 289.

78. M. Daffé, A Laszlo, H.L. David, J. Gen. Microbiol., 1989, 135, 2759.

79. F. Papa, M. Luquin, D.H. David, Res. Microbiol., 1992, 143, 327.

80. D.E. Minnikin, R.C. Bolton, M. Magnusson, FEMS Microbiol. Lett., 1990, 67, 55.

81. M. Daffé, M. McNeil, P.J. Brennan, Biochemistry, 1991, 30, 378.

82. D.E. Minnikin, G. Dobson, D. Sesardic, M. Ridell, J. Gen. Microbiol., 1985, 131, 1369.

83. D.E. Minnikin, M. Ridell, G. Wallerström, G.S.
     Besra, J.H. Parlett, R.C. Bolton, M. Magnusson,
     Acta Lepr., 1989, 7 (Suppl.1), 51.
84. D.E. Minnikin, M. Ridell, J.H. Parlett, R.C.
     Bolton, FEMS Microbiol. Lett., 1987, 48, 175.
85. A. Lemassu, M.A. Lanéelle. M. Daffé, FEMS
     Microbiol. Lett., 1991, 78, 171.
86. M. Ridell, G. Wallerström, D.E. Minnikin, R.C.
     Bolton, M. Magnusson, Tubercle and Lung Diseases,
     1992, 72, 101.
87. G.S. Besra, R.C. Bolton, M. McNeil, M. Ridell,
     K.E. Simpson, J. Glushka, H. van Halbeck, P.J.
     Brennan, D.E. Minnikin, Biochemistry, in press.
88. T. Fujiwara, Agric. Biol. Chem., 1991, 55, 2123.
89. P. Cruaud, F. Papa, H.L. David, M. Daffé,
     Acta Lepr., 1989, 7 (Suppl.1), S94.
90. J.T. Belisle, P.J. Brennan, J. Bacteriol.,1989,
     171, 3465.
91. G. Dobson, D.E. Minnikin, S.M. Minnikin, J.H.
     Parlett, M. Goodfellow, M. Ridell, M. Magnusson in
     'Chemical Methods in Bacterial Systematics', Ed.
     M. Goodfellow, D.E. Minnikin, Academic Press,
     London, 1985, p.237 .
92. D.E.Minnikin, R.C. Bolton, G. Dobson, A.I. Mallet,
     Proc. Jap. Soc. Med. Mass Spectrom., 1987, 12, 23.
93. D.E. Minnikin,G. Dobson, K. Venkatesan, A.K.
     Datta, J.H. Parlett, R.C. Bolton, Health
     Cooperation Papers, 1986, 7, 211.
94. D.E. Minnikin, G.S. Besra, A.K. Datta, A,I.
     Mallet, R.C. Bolton, Health Cooperation Papers,
     1992, 12, 53.

Biodegradable Surfactants Derived from Phenolic Lipids

J. H. P. Tyman and I. E. Bruce

DEPARTMENT OF CHEMISTRY, BRUNEL UNIVERSITY, UXBRIDGE,
MIDDLESEX UB8 3PH, UK

1   INTRODUCTION

The phenolic lipids comprise two main groups, those of polyketide origin such as anacardic acid (1) which occurs in <u>Anacardium occidentale</u> or those from an isoprenoid pathway of which α-tocopherol (2) is a well-known example.

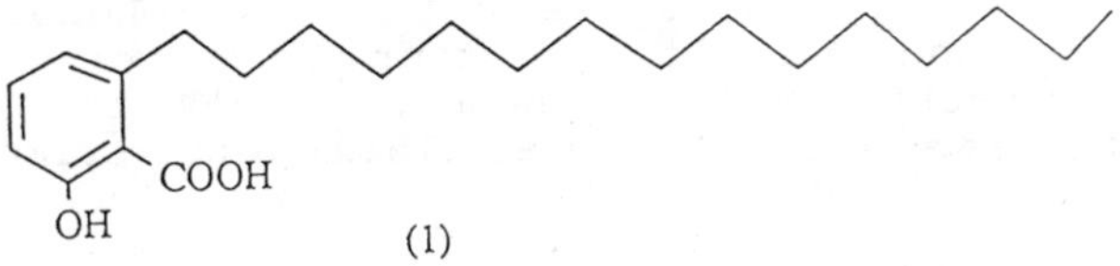

(1)

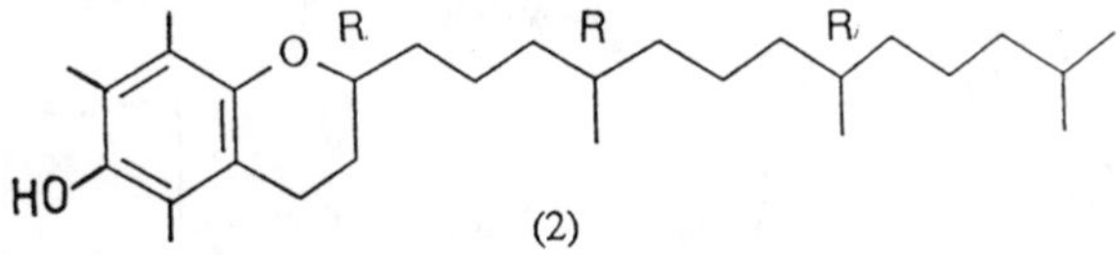

(2)

There are now recognised to be many botanical and biological species in the former group which contain monohydric, dihydric phenols and phenolic acids[1] and members of this whole class may be viewed as fatty acids, with methylene group-interrupted conjugation and a phenolic head group in place of the usual carboxyl.   The cashew tree, <u>Anacardium occidentale</u>, from the <u>Anacardiaceae</u> is the most widely distributed type and a source of the phenolic lipid, cashew nut-shell liquid (technical CNSL) a by-product of industrial processing of the cashew nut which is primarily directed to production of the kernel.   It has been suggested[2] that by the turn of the century cashew nut production could lie between 5×10^5 and 10^6 tonnes p.a. Originally indigenous to Brazil, the cashew tree is grown in South India, East Africa other equatorial and subequatorial regions and more recently in Thailand and Indonesia.

<u>Industrial Processing of the Cashew</u>

The naturally occurring phenolic lipids in the cashew are present in the shell of the nut and comprise anacardic acid (1) and cardol (3) and in the major industrial processing method used (Sturtevant Process)- other techniques such as the Oltremare process are also available-cashew nuts are roasted at 180-190° C in a bath of hot technical CNSL whereby cardanol (4) results by hot decarboxylation. Cardanol is only a minor component of the natural product and is in reality a semi-synthetic material.

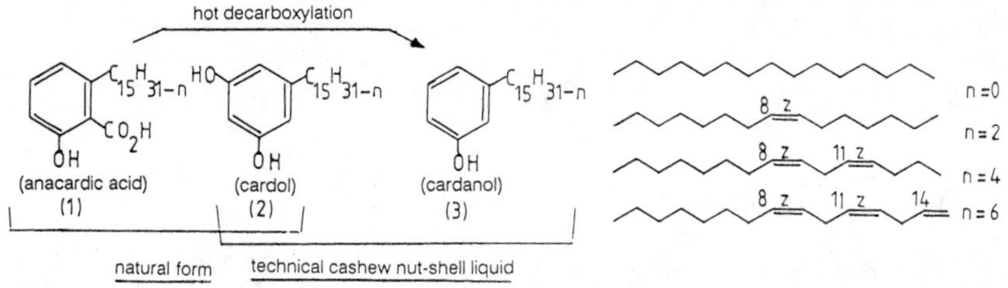

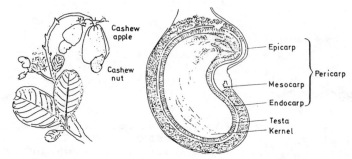

Fig. 1 Phenolic Lipids in/from <u>Anacardium occidentale</u>

Fig. 1 gives the formulae of the three phenolic lipids referred to and details the four types of side-chain present in each component phenol, namely the saturated (15:0), monoene (15:1), diene (15:2) and triene (15:3) constituents.

Fig. 2a,b   The Cashew apple and cross section of the nut

The unusual structure of the cashew nut is shown in Fig.2b, (the average length is 2-3 cm., and the weight, 3-5g.) 2a showing the external nut attached to the cashew apple.

Of the total raw nut weight, the kernel is approximately 20-25%, natural CNSL in the shell, 18-27% and the remainder consists of the shell material and the testa.

Industrial processing affords approximately 40% of technical CNSL from the available natural CNSL, representing about 10% of the weight of the raw nut. Ways have been found to obtain almost the theoretically available technical CNSL through solvent extraction of the mechanically cut and fragmented nut shell followed by catalytic decarboxylation[3].

In the hot decarboxylation the outer epicarp is burst by the CO_2 pressure liberating CNSL from the honey-combed mesocarp, the inner shell, the endocarp, still being intact and together with the testa skin completely affording protection of the kernel from the CNSL. In a subsequent process stage the inner endocarp shell is broken by a novel aerodynamic/shattering technique in which the highly-prized kernel survives mostly intact. The kernel itself is a conventional nut containing, as is normal, its own triglyceride oil.

The biodegradable surfactants described in the present work were obtained from cardanol and cardol. The former is the major component of technical CNSL, usually present at ₋70-75% together with cardol (₋15-20%), minor components and polymeric material.

2 SEPARATION OF CARDANOL FROM TECHNICAL CNSL

The main current use of technical CNSL is in the manufacture of friction dusts for which the dark brown colour of the raw material and the presence of dimeric/polymeric material is immaterial. However in chemical applications purification of the CNSL and the separation of cardanol is invariably desirable. Cardanol so obtained is a mixed product consisting of (15:0), (15:1), (15:2), and (15:3) constituents the separation of which for industrial utilisation is not practical. Accordingly the objective of separation processes is directed to the removal of the dihydric phenol cardol, (probably the main cause of the dark colour), other minor components and the polymeric material and to achieve this, if possible, without affecting the existing proportions of unsaturated constituents. The separation of the phenolic lipids in technical CNSL on a preparative scale may be effected by chromatography, and on a larger scale either by molecular distillation or by chemical methods combined with a final conventional high vacuum distillation to remove traces of colour and polymeric material.

Chromatographic Analysis of Technical CNSL. Prior to the establishment of a separatory process an effective analytical method is essential to enable the proportion of all the materials referred to to be readily determined in a single run. HPLC analysis by the reversed-phase method with gradient elution and an internal standard enables all the constituents and the polymeric material to be quantitatively determined[4].

Fig. 3a shows the composition of a sample of technical CNSL containing a relatively high proportion of polymer, P (21.6%) together with cardanol triene, F (22.5%), cardanol diene, E (12.5%), cardanol monoene, D (26.1%), saturated cardanol, S (3.1%), cardol triene, I (6.6%), cardol diene, (2.6%), cardol monoene, G (1.1%) and minor materials (3.9%). The total 'mixed' cardanol in this sample was 63.7% although in better quality material it lies within the range 70-75%. In freshly distilled technical CNSL, fig. 3b, the polymer is virtually absent. Because technical CNSL is obtained by thermal processing towards

which conditions all the diene and triene constituents are
susceptible to polymerisation and degradation[5] the

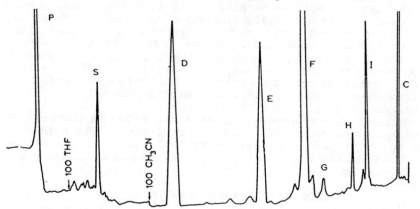

Fig 3a HPLC separation of old CNSL on Spherisorb ODS (5 μm)
by gradient elution, starting with acetonitrile
water (66:34). Flow-rate 1.7 ml/min. P = Polymer;
C = p-cresol

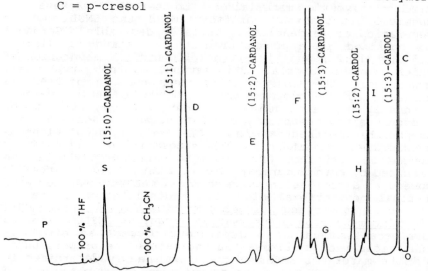

Fig 3b HPLC separation of distilled CNSL on Spherisorb ODS
(5 μm) by gradient elution, starting with
acetonitrile water (66:34). Flow-rate. 1.7 ml/min. Peaks
D I and S as in Fig. 3a: P = polymer: C = p-cresol.

assessment of the % polymer is preferable by a 'cold
method' such as HPLC rather than by GLC. Chromatographic
methods for the analysis of anacardic acids and
accompanying phenolic lipids have been reviewed[6].

Chromatographic Separation of Cardanol from Technical
CNSL. Cardanol has been separated from technical CNSL by
liquid chromatography in columns containing Silica Gel H
(Type 60) with solute/absorbent ratios in the range 1:5-1:6
by step-wise elution with the solvents light petroleum (60-

80° C) and diethyl ether. It was found that the column could be re-used four times with batches of 30g of solute[7]. Although macro batches up to 250g have been processed there is little doubt that distillation methods or chemical purification/high vacuum distillation techniques are preferable for larger scale work.

Separation by Molecular Distillation. Molecular distillation with a multi-stage still provides the only satisfactory distillation method for the separation of cardanol and cardol[8] (no separation of the three unsaturated constituents of cardanol was observed). Conventional fractional distillation and spinning band techniques are ineffective since the prolonged heating to establish equilibrium in the refluxing process merely causes extensive polymerisation and a poor throughput. With a 10-stage laboratory rotary molecular still technical CNSL afforded a yield of approximately 60% of almost colourless cardanol completely free from cardol. It is understood that comparatively large scale distillations of Brazilian technical CNSL have been effected in that country. By the use of wide-bore equipment, short-path conventional stills together with oil-bath heating, excellent recoveries of products comparatively free of colour and polymeric material although with minimal removal of cardol, may be made by high vacuum distillation (in which the temperature is referably kept below 180° C).

Organic Base Addition and High Vacuum Distillation. Combined chemical purification/vacuum distillation has been found effective for partially reducing the % cardol in cardol[9,10]. A selected amine added to technical CNSL resulted in preferential reaction with the more acidic cardol and upon vacuum distillation of the mixture purer cardanol was recovered. Typically, technical CNSL containing cardanol (86.9%), cardol (10.3%), and 2-methylcardol (2.3%) upon treatment with diethylenetriamine (0.5 mole) and reaction at ambient temperature for 24 hours gave by vacuum distillation a 62.3% recovery of a product containing cardanol (97.7%) and cardol (2.0%).

Selective Mannich Reaction of Cardol and Vacuum distillation. The selective reaction of cardol in preference to cardanol under Mannich reaction conditions with diethylenetriamine or 4-aminobutane and aqueous formaldehyde in methanolic solution resulted in the separation of cardol as a low polymeric material insoluble in the solution[11,12,13]. Technical CNSL (1mol) with 40% aqueous formaldehyde (1.2mol, CH_2O) and diethylenetriamine (0.125mol) in methanol (1250cm^3) afforded after 30 mins. a dark methanol-insoluble lower layer and an upper lighter phase which after solvent recovery and work-up gave crude cardanol (63%) containing a trace of cardol, a small amount of 2-methylcardol and some polymeric and coloured material. High vacuum distillation furnished cardanol containing traces of 2-methylcardol since this component only partially undergoes the Mannich reaction.

Phase Separation of Technical CNSL to give Cardanol and Cardol. The desirability of having a non-distillation method for larger scale use which would lead to cardanol free from all the other component phenols and permit the recovery of cardol itself rather than its loss through conversion to a Mannich base led to studies of the phase separation of Technical CNSL[14]. In the non-aqueous immiscible bi-phase solvent system of light petroleum and an alkane diol, cardanol specifically enters the upper hydrocarbon phase and cardol with 2-methylcardol is found in the lower diol phase. In this way a highly selective separation can be achieved particularly with butane-1,4- and pentane-1,5-diols.

Table 1. Phase Separation to Technical CNSL in Non-aqueous Light petroleum/Diol systems (% cardanol and cardol by HPLC and TLC analysis)

Diol	Cardanol (% yield in l.p. phase)	Analysis of cardanol %cardanol	cardol
Ethane-1,2-	96	9.95	5.04
Propane-1,2-	90	97.91	1.87
Propane-1,3-	89	97.14	2.85
Butane-2,3-	54	98.47	1.53
Butane-1,3-	46	98.74	1.25
Butane-1,4-	84	99.08	0.92
Pentane-1,5-	93	99.81	0.19
Digol	41	94.83	5.16
Polygol	7	98.56	1.43

Table 1 shows the results obtained with a number of readily available diols. Polymeric material emerges with the cardanol and coloured material is found with the cardol. Work-up and solvent recovery of the two separated phases gives cardanol in a pure form and cardol also available for chemical utilisation.

3   THE SYNTHESIS OF SURFACTANTS FROM CARDANOL AND CARDOL

The availability of improved separational processes for obtaining cardanol and cardol has enabled them to be used in applications where the long alkyl chain is advantageous namely in the semi-synthesis of surfactants. Anionic derivatives have been prepared by the reaction of mixed cardanol (that is material containing triene, diene, monoene and saturated constituents) with sulphating and sulphonating reagents[15]. Hydrogenated cardanol has been sulphonated[16,17], and the surface properties of the anionic product, 3-pentadecylphenol-4-sulphonic acid and its salts studied. By the Mannich reaction[18] 3-pentadecyl-6-dimethylaminophenol and the mixed unsaturated analogues have been obtained and the saturated (15:0) cationic methosulphate resulting from quaternisation with dimethyl sulphate examined for its germicidal properties[19,20]. The NMR spectra of cardanol ethoxylate and its (15:0) version have been studied[21] although the relative and general properties of the group had not been examined and it was deemed from our work with the anionic and cationic series that an unsaturated product would be more usefully soluble

than the saturated analogue. The ethoxylation of
nonylphenol derived from petrochemical sources has long
been established industrially and both the ethoxylate and
its sulphate are produced on a large scale. By contrast,
the cashew phenols represent relatively cheap semi-
synthetic materials from a natural resource. It seemed
highly probable that such biosynthesised substances (some
cardanol as indicated earlier is present in natural CNSL)
could also be more biodegradable than their petrochemical
relatives and on these tenuous grounds the overall
properties of cardanol and cardol ethoxylates should be
worth examination.

The Preparation of Cardanol and Cardol Polyethoxylates

Phenol ethoxylates can be synthesised with acidic or
base catalysis at elevated temperatures, conditions which
can, if prolonged with cardanol or cardol, result in either
polymerisation or this accompanied by isomerisation.
Cardanol and cardol were ethoxylated[11] by essentially the
same procedure in equipment based on a modification of that
described[22]. Under a nitrogen atmosphere a stirred mixture
of mixed cardanol containing 0.5% potassium hydroxide or
sodium hydroxide was heated for 30mins. at 180° C (for
cardol the temperature was 160° C) after which the nitrogen
flow was replaced by that of ethylene oxide. The reaction
mixture was sampled every 30 mins. under nitrogen and the
procedure continued for 8 hours giving a sequence of 16
samples of polyethoxylated cardanol which were then
analysed, after neutralisation of the base catalyst, by
HPLC and by NMR. The average number of ethylene oxide
units n, per cardanol or cardol molecule had values ranging
from 1 to 50. Nonylphenol (Aldrich Chemicals) in which the
side chain consists of a number of isomeric structures was
ethoxylated as a reference compound.

The Mechanism of Reaction. The mechanism of the
polyethoxylation reaction consists of nucleophilic attack
of the phenolate anion on ethylene oxide resulting in ring
opening with formation of an alcoholate anion which then
reacts further with ethylene oxide leading to progressive
development of the polyethoxylate chain to give n units
(the first stage is shown in Scheme 1), in which R
represents a 3-substituted $C_{15}H_{31-n}$ or a 4-substituted tert-
C_9H_{19} group.

Scheme 1

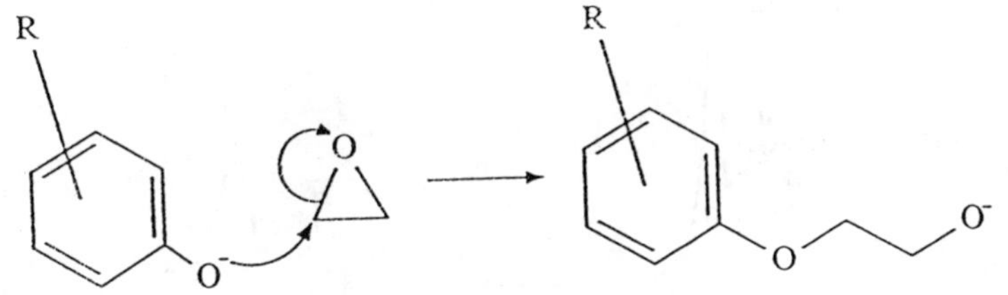

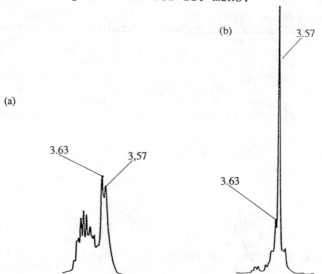

A competing reaction can occur in the early stages in which the anion formed from ethylene oxide and the hydroxide ion can react further to give the polyethylene glycol anion. However as the alcoholate anion from reaction of the phenolate anion accumulates, this side reaction soon diminishes.

Characterisation of Polyethoxylates from Phenolic Lipids

Products obtained from the polyethoxylation of cardanol, of cardol and other phenolic lipids were analysed by ^{1}H NMR spectroscopy, HPLC analysis and independent synthesis of the first six pure members of the series (n=1 to 6) to enable the oligomers revealed by chromatographic analysis to be identified and averaged ethoxylate numbers to be calculated.

^{1}H NMR Spectroscopy. NMR analysis of polyethoxylated cardanol as a 1% solution in deuteriochloroform at 90MHz readily enabled the extent of the reaction to be determined.

Fig.4 shows (a) the peaks of the methyleneoxy group in polyethylene glycol at 3.63δ, after 30mins. and in cardanol polyethoxylate at 3.57δ, and (b) at the same chemical shifts, following reaction for 210 mins.

(b)   3.57

(a)

3.63   3,57

3.63

Fig.4a,b ^{1}H NMR of CH$_2$O groups in cardanol polyethoxylate

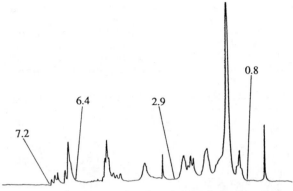

Fig.5 ^{1}H NMR of mixed cardanol

Fig.5 shows the NMR spectrum of mixed cardanol with aliphatic protons (CH$_2$, CH$_3$) from 0.8 to 2.9ppm, olefinic protons between 4.8 and 6.0ppm, and four aromatic protons from 6.4 to 7.2ppm. In the NMR spectrum of cardanol polyethoxylate the above peaks are greatly diminished due to the preponderance of the new multiplet from 3.2 to 4.3 ppm for the methyleneoxy group (the phenolic hydroxyl proton is also in this region). Some reduction in the olefinic peaks, particularly of the terminal vinyl group is attributable to base isomerisation. From the integration for the methyleneoxy peak (between 3.2 and 4.3 ppm) in relation to that for the four proton aromatic peak (between 6.4 and 7.2ppm) the number of oxyethoxy protons was calculated, allowance being made for the phenolic hydroxyl, and hence the averaged number of oxyethoxy groups found for each sample withdrawn from the reaction mixture.

Time (mins)	Average number of EO units/molecule	Time (mins)	Average number of EO nits/molecule
30	0.6	270	18.3
60	1.0	300	21.1
90	1.8	330	25.9
120	5.5	360	28.9
150	7.2	390	32.6
180	10.7	420	37.2
210	13.5	450	42.2
240	16.9	480	48.0

Table 2. Time of reaction of cardanol with ethylene oxide and the average EO units/mole (calculated by NMR)

Table 2 gives the results calculated for mixed

cardanol for samples taken at 30 minute intervals up to 8 hours. Similar results were shown by 3-pentadecylphenol (saturated cardanol) although the reaction appeared significantly slower until 75% completed while mixed cardol exhibited throughout a reduced rate which resulted in the final product containing an averaged 34.7 ethylene oxide units after 8 hours.

HPLC Analysis of Polyethoxylates of Phenolic Lipids. HPLC analysis is particularly suitable for the analysis of the oligomeric mixtures (all of which possess a Poisson distribution of components) resulting from polyethoxylations of phenolic lipids since with TLC, quantification is difficult, a limited number of the oligomers are eluted from GLC columns even after derivatisation and SEC (size exclusion chromatography) is less selective than HPLC operated with gradient elution. The adsorption mode with gradient elution and a Bondpack semipolar NH_2 propyl column has been employed for octylphenol polyethoxylates[23] while for dodecylphenol and dodecanol polyethoxylates, LiChrosorb DIOL columns[24] have afforded less resolved chromatograms. The former system was used for cardanol, cardol and 3-pentadecyl polyethoxylates with a 250x4.6 mm. column, gradient elution based on solvent A (20% THF in n-hexane) and B (10% water in iso-propanol), a flow rate of 1ml/min. an ideal loading of 200μg in 20μl diethyl ether and with UV detection[11].

A typical HPLC chromatogram for cardanol polyethoxylate obtained under these analytical conditions is shown in fig.6. and fig.7 illustrates the result of 'spiking' the mixture with synthetic compounds obtained as described in the next section.

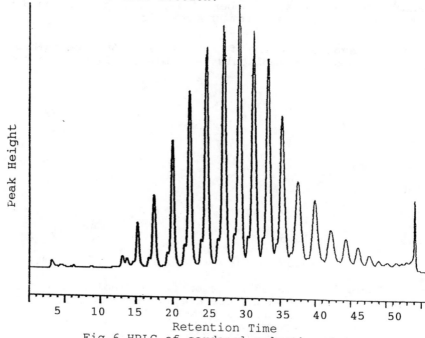

Fig.6 HPLC of cardanol polyethoxylate

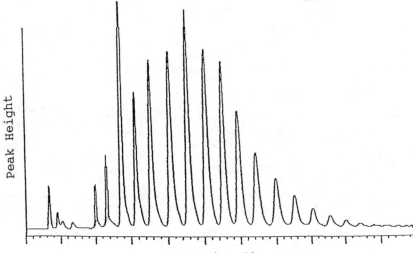

Retention Time
Fig.7 HPLC of 'spiked' Cardanol polyethoxylate

Retention time (min)	Area %	Ethoxylate number
15.24	1.24	2
17.47	2.63	3
20.01	4.59	4
22.36	7.76	5
24.64	10.23	6
26.99	11.56	7
29.14	12.79	8
31.32	11.97	9
33.45	10.65	10
35.45	8.29	11
37.73	5.41	12
40.16	3.83	13
42.45	2.18	14
44.59	1.35	15
46.30	0.81	16

Table 3   HPLC Retention Time, % Peak Area and EO number for Fig.6

In the HPLC method the saturated, monoene, diene and triene constituents of each oligomer elute as a single peak. The cardanol used for the polyethoxylate displayed in fig.6 had been prepared by the Mannich base method and the small peaks which precede each cardanol ethoxylate oligomer are due to traces of 2-methylcardol present in the starting material. Table 3 gives for cardanol polyethoxylate the numerical data on retention times, % peak area and the number of ethoxylate units shown in fig.6. The average ethoxylate number for each reaction sample which was calculated from NMR spectral data as given in Table 2 can be derived also from the % peak area found by HPLC through the incorporation of a known weight of a reference synthetic oligomer used as a 'spike' in the cardanol polyethoxylate.

In this general way, weight-average ethoxylate numbers and hence molar-average ethoxylate numbers can be deduced[25], on the accepted basis that UV molar extinction coefficients are dependent on the phenolic absorption and independent of the ethoxylate chain.

Time (min)	Weight-average Ethoxylate Number	Molar-average Ethoxylate Number	Average Number of EO units/mol (nmr)
30	0.93	0.62	0.6
60	1.47	1.14	1.0
90	2.14	1.86	1.8
120	6.14	5.71	5.5
150	8.38	7.52	7.2
180	11.62	10.98	10.7
210	14.21	13.44	13.5
240	17.59	16.98	16.9
270	19.67	18.55	18.3
300	22.61	21.33	21.1
330	27.72	27.24	25.9
360	30.11	28.47	28.2
390	35.16	33.02	32.6
420	39.27	37.17	37.2
450	44.56	42.37	42.2
480	50.17	48.51	48.0

Table 4 Reaction Time, ethoxylate numbers (HPLC and NMR)

Table 4 gives for each sample withdrawn at 30 minute intervals in the formation of cardanol polyethoxylate, the weight-average and the molar-average ethoxylate numbers determined from HPLC data, with for comparison the values obtained by the NMR procedure (taken from Table 3). Similar procedures were adopted for the polyethoxylates from cardol and from 3-pentadecylphenol and in all three cases good agreement was observed.

<u>Synthesis of Individual Oligomers</u>. As pointed out in the previous section it was essential to correlate HPLC retention data with ethoxylate chain length in order to find the average ethoxylate number since the surfactant efficiency was dependent upon this profile. The first six members of the oligomers (n=1 to 6) were synthesised for the 3-pentadecyl (5) and the mixed cardanol (6) series. It was hoped that this could be achieved by the phase transfer method[26] and indeed that this technique might be applicable to the ethoxylation reaction itself since it had been successful for phenolic ethers in the cardanol, cardol and anacardic acid series[27]. However this methodology was abandoned in favour of standard nucleophilic reactions by a modification of a described procedure[28] in which it proved essential to form the phenoxide anion of cardanol with sodium metal for formation of the first member (n=1) and with potassium metal for reaction of the first member to yield the higher members (n=2 to 6) in each series as illustrated in Scheme 2. For the first compound the halide was 2-bromoethanol. This compound was also derived by reaction of the phenoxide with ethyl bromoacetate, hydrolysis of the product to the corresponding acid which was reduced with lithium aluminium hydride. For the higher members a series of chlorohydrins without protection of the hydroxyl group were used. All compounds were examined spectroscopically and, where relevant, by elementary analysis. The yields of the compounds are indicated below the formulae.

Scheme 2

$$C_{15}E_0 \xrightarrow{\ E_2\ } C_{15}E_2 \qquad C_{15}E_3 \xrightarrow{\ E_2\ } C_{15}E_5$$

$$C_{15}E_0 \xrightarrow{\ E_3\ } C_{15}E_3 \qquad C_{15}E_3 \xrightarrow{\ E_3\ } C_{15}E_6$$

$$C_{15}E_1 \xrightarrow{\ E_3\ } C_{15}E_4$$

C_{15} = Pentadecyl phenol

E_x = Number of ethylene oxide units

n = 1
n = 2 (45%)
n = 3 (46%)
n = 4 (36%)
n = 5 (42%)
n = 6 (46%)

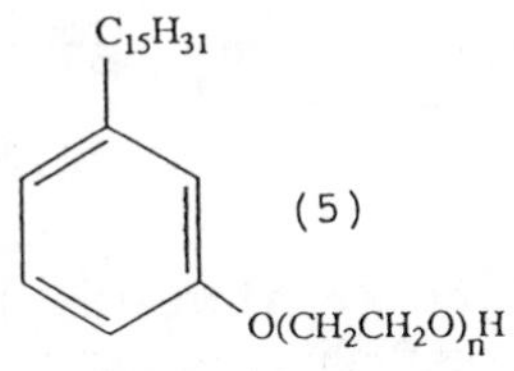

$C_{15}H_{31}$

(5)

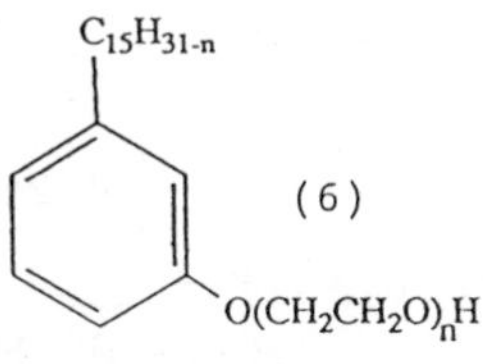

$O(CH_2CH_2O)_nH$

$C_{15}H_{31-n}$

(6)

$O(CH_2CH_2O)_nH$

n = 1 (63%)
n = 2 (42%)
n = 3 (43%)
n = 4 (38%)
n = 5 (40%)
n = 6 (47%)

4 SOME SURFACE PROPERTIES OF POLYETHOXYLATES OF PHENOLIC LIPIDS

Surfactancy may be judged by a number of properties such as foaming ability, reduction in surface tension, solubilisation action and other characteristics which are often examined on formulated products. In the present approach the reduction of the surface tension was determined by the du Nouy method at 24° C, relative to a standard of distilled water, on 1% aqueous solutions of each of the 16 samples from cardanol, cardol and 3-pentadecylphenol polyethoxylates. The results for the three groups are shown in Tables 5,6 and 7 and are

Av Number of EO groups	Reduction in Surface Tension	Av. Number of EO groups	Reduction in Surface Tension
0.62	4.6	18.55	28.6
1.14	7.9	21.33	26.0
1.86	8.9	26.24	24.5
5.71	16.4	28.47	23.5
7.52	24.0	33.02	23.0
10.98	31.4	37.17	22.4
13.44	34.2	42.37	21.7
16.98	30.8	48.51	20.2

Table 5 Reduction in Surface Tension (Nm^{-1}), for cardanol polyethoxylate with respect to water

Av. Number of EO groups	Reduction in Surface Tension	Av. Number of EO groups	Reduction in Surface Tension
0.27	1.4	11.92	23.6
0.66	2.1	15.22	22.1
0.91	2.9	17.81	21.7
2.38	7.4	19.28	20.3
3.21	11.1	24.21	18.9
4.87	17.5	25.17	17.8
6.89	19.0	29.64	16.3
10.31	24.2	36.49	15.1

Table 6 Reduction in Surface Tension (Nm^{-1}), for cardol polyethoxylate

Av. Number of EO groups	Reduction in Surface Tension	Av. Number of EO groups	Reduction in Surface Tension
0.71	3.3	10.12	28.4
1.28	4.2	14.81	33.9
1.82	5.2	18.66	29.5
1.96	8.3	24.29	25.2
2.03	8.5	27.13	24.2
2.74	10.0	34.61	22.8
4.39	15.2	38.68	22.0
6.21	19.9	47.22	20.9

Table 7 Reduction in Surface Tension (Nm^{-1}), for 3-Pentadecylphenol polyethoxylate.

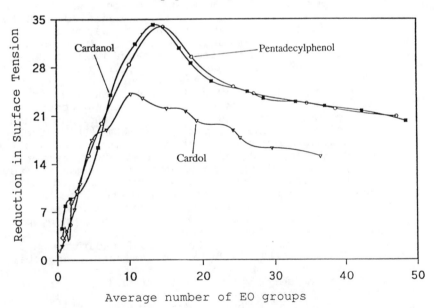

Fig. 8 Variation in Surface Tension with ethylene oxide chain length for cardanol, cardol and 3-pentadecyl polyethoxylates.

illustrated graphically in Fig.8 which shows the variation in surface tension reduction with ethylene oxide chain length for these materials.

Mixed cardanol polyethoxylate containing an average number of 13.4 ethyloxy groups, 3-pentadecylphenol polyethoxylate with 14.8 and cardol polyethoxylate with 10.3 exhibited the greatest reductions in surface tension and the best surfactant properties would be expected to be revealed at these levels of ethoxylation. A 1% solution of nonylphenol polyethoxylate (reduction in surface tension, 33 Nm^{-1}) with an average of 8 ethylene oxide units offered a comparison. From these results it appears desirable to have low levels or no cardol polyethoxylate present in the cardanol analogue.

5   THE BIODEGRADATION   OF PHENOLIC LIPID POLY ETHOXYLATES

The importance of nonionic surfactants for detergent formulations has grown considerably in recent years and increasing attention has to be paid to their environmental compatibility and for the requirement to furnish proof of their biodegradability under environmental conditions. Nevertheless although the whole area of biodegradability has been reviewed[29,30], relatively little work has appeared on the properties of nonionic compared with anionic surfactants. In the present work[11], a modified OECD test which is designed to screen substances, was employed for the determination of biodegradability. It uses a low concentration of microorganisms and has a maximum duration of 28 days. Substances passing this test (> 70% DOC, dissolved organic carbon, removal) can be assumed to be extensively degraded in sewage treatment and are likely to be rapidly removed from bulk water. In the procedure used the test compound was dissolved in an inorganic medium (concentration of the compound 40-100mg/1) which was inoculated with a low concentration of microorganisms from a mixed population and aerated at a constant temperature for 28 days during which time biodegradation was monitored by DOC analysis. The procedure was checked with a reference compound of known biodegradability such as glucose and a comparison of the phenolic lipid polyethoxylates made with a known product, nonylphenol polyethoxylate. The mixed cardanol and cardol polyethoxylates used for the biodegradation studies contained 13.4 and 10.3 ethylene oxide units respectively and a direct comparison of the saturated and polyunsaturated series was made by using hydrogenated versions of the two preceeding compounds. The dichromate oxidation procedure was employed to determine the levels of dissolve organic carbon in the samples (DOC) by trapping of the evolved carbon dioxide in aqueous barium hydroxide and acidic titration[31]. The results of these biodegradation experiments are shown in Table 8 for cardol polyethoxylate (10.5EO), its saturated analogue, 3-pentadecylphenol polyethoxylate (13.4EO), and nonylphenol polyethoxylate (8EO). The results are shown graphically for cardanol, pentadecylphenol and nonylphenol polyethoxylates if Fig. 9 and for cardol and saturated cardol polyethoxylates with a glucose reference in Fig.10.

Time (days)	Cardol 10.5EO	Saturated cardol 10.5EO	Pentadecyl phenol 13.4EO	Nonylphenol 8EO
0	100	100	100	100
4	86	90	67	96
8	74	76	44	90
12	61	77	34	91
16	54	57	29	81
20	49	51	34	81
24	41	47	27	79
28	37	46	25	77

Table 8   Biodegradation of Cardol, Saturated Cardol, 3-Pentadecylphenol and Nonylphenol polyethoxylates as monitored by DOC Analysis.

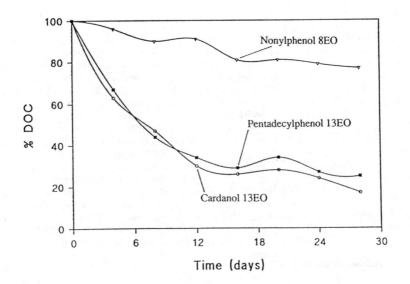

Fig.9   Biodegradation of Cardanol, Pentadecylphenol and Nonylphenol polyethoxylates.

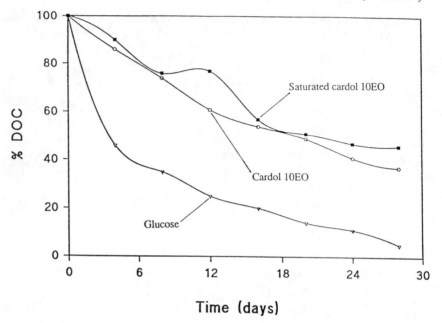

Fig.10   Biodegradation of Cardol and Saturated Cardol
             Polyethoxylates and Glucose.

CONCLUSIONS

The major phenolic lipids in technical CNSL, cardanol
and cardol have been reacted with ethylene oxide by base
catalysis to yield nonionic surfactants.   These products
have also been hydrogenated to give their pentadecyl
equivalents.   All these products are mixtures of oligomers
the surfactant properties of which were studied with regard
to surface tension measurements.   From these, cardanol
polyethoxylate and cardol polyethoxylate with an average of
13.4 and 10.5 ethylene oxide units respectively afforded
the greatest reduction in surface tension.   In a modified
'shake flask' method for biodegradability monitored by the
DOC procedure they were found to be highly biodegradable
compared with nonylphenol polyethoxylate.   The cardanol
compound possessed the highest level of surfactant and
biodegradability properties and it appears useful to remove
cardol from the raw material used for the synthesis.   A
number of methods have been devised for this purpose.

ACKNOWLEDGEMENTS

We are grateful to 3M Research Ltd., and S.C. Johnson
Ltd. for financial assistance and to Dr. G. Hart of the
latter organisation for making some facilities available.

REFERENCES

1.   J.H.P. Tyman, 'Studies in Natural Products Chemistry',
     Ed. Atta-ur-Rahman, Elsevier, Amsterdam, 1991, Vol. 9,

p. 313.
2. J.G. Ohler, 'Cashew', The Royal Tropical Institute, Amsterdam, 1979, Chapter 2, p. 21.
3. J.H.P. Tyman, R.A. Johnson, M. Muir, and R. Rokhgar, J. Am. Oil Chem. Soc., 1989, 66, 553.
4. J.H.P. Tyman, V. Tychopoulos and B.A. Colenutt, J. Chromatography, 1981, 213, 287.
5. J.H.P. Tyman and V. Tychopoulos, J. Sci. Food Agric. 1990, 52, 71.
6. J.H.P. Tyman, 'Chromatography of Anacardic Acids', in CRC Handbook of Chromatography: Lipids Vol. III, Eds., H.K. Mangold, K.D. Mukherjee and N. Weber, 1992, CRC Press, Boca Raton, Florida (in press).
7. S.K. Sood, J.H.P. Tyman, A.A. Durrani and R.A. Johnson, Lipids, 1986, 21, 241.
8. A.A. Durrani, G.L. Davis, S.K. Sood, V. Tychopoulos and J.H.P. Tyman, J. Chem. Tech. Biotechnol., 1982, 32, 681.
9. M. Patel, J.H.P. Tyman and A. Manzara, U.K. Pat. Appln., 8100208.
10. M. Patel, M. Phil. Thesis, Brunel University, 1979.
11. I.E. Bruce, Ph.D. Thesis, Brunel University, 1991.
12. J.H.P. Tyman, U.K. Pat. Appln., GB 2152925A.
13. V. Tychopoulos, Ph.D., Thesis, Brunel University, 1983.
14. J.H.P. Tyman, U.K. Pat. Appln., M91/5038/GB.
15. J.H.P. Tyman and A.A. Durrani, (unpublished work).
16. S.C. Sethi, B.C. Subba Rao and S.B. Kulkarni, Indian J. Technol., 1963, 1, 348.
17. D. Wasserman, U.S.P., 2586191.
18. J.H.P. Tyman and V. Tychopoulos, Synth. Commun., 1986, 16, 1402.
19. A.S. Gulati, V.S. Krishnamchar and B.C. Subba Rao, Indian J. Chem., 1964, 2, 114.
20. J.H.P. Tyman, Chem. Soc. Rev., 1979, 8, 499.
21. R. Subbarao and V.P. Harigopal, Fette, Seifen and Anstrichm., 1975, 77, 197.
22. J.V. Karabinos, G.E. Bartels and G.E. Rapella, J. Amer. Oil Chem. Soc., 1954, 31, 419.
23. F.P.B. Van der Maeden, M.E.F. Biemond and P.C.G.M. Janssen, J. Chromatogr., 1978, 149, 539.
24. L. Zeman, J. Chromatogr., 1986, 363, 233.
25. A. Nozawa and T. Ohnuma, J. Chromatogr. 1980, 187, 261.
26. A. McKillop, J.-C. Fiaud and R.P. Hug, Tetrahedron, 1974, 30, 1379.
27. J.H.P. Tyman and S.K. Lam, J. Chem. Soc. Perkin I, 1981, 1942.
28. Y. Abe and S. Watanabe, Fette Seifen Anstrich., 1972, 9, 534.
29. R.D. Swisher, J. Amer. Oil Chem. Soc., 1963, 40, 648.
30. R.D. Swisher, 'Surfactant Biodegradation', M. Dekker, New York, 1987, 2nd Edn.
31. S.J. Huddleston, and C. Alldred, J. Amer. Oil Chem. Soc., 1964, 41, 723.

Subject Index